Asian Scientists on the Move

The growing scientific research output from Asia has been making headlines since the start of the twenty-first century. But behind this science story, there is a migration story. The elite scientists who are pursuing cutting-edge research in Asia are rarely "*homegrown*" talent but were typically born in Asia, trained in the West, and then returned to work in Asia. *Asian Scientists on the Move* explores why more and more Asian scientists are choosing to return to Asia, and what happens after their return, when these scientists set up labs in Asia and start training the next generation of Asian scientists. Drawing on evocative firsthand accounts from 119 Western-trained Asian scientists about their migration decisions and experiences, and in-depth analysis of the scientific field in four country case studies – China, India, Singapore and Taiwan – the book reveals the growing complexity of the Asian scientist migration system.

ANJU MARY PAUL is an international migration scholar with a research focus on emergent migration patterns to, from and within Asia. Her previous books include the award-winning *Multinational Maids: Stepwise Migration in a Global Labor Market* (Cambridge University Press 2017) and *Local Encounters in a Global City* (Ethos Books 2017).

Asian Scientists on the Move

Changing Science in a Changing Asia

ANJU MARY PAUL
Yale-NUS College

CAMBRIDGE
UNIVERSITY PRESS

University Printing House, Cambridge CB2 8BS, United Kingdom

One Liberty Plaza, 20th Floor, New York, NY 10006, USA

477 Williamstown Road, Port Melbourne, VIC 3207, Australia

314–321, 3rd Floor, Plot 3, Splendor Forum, Jasola District Centre, New Delhi – 110025, India

103 Penang Road, #05–06/07, Visioncrest Commercial, Singapore 238467

Cambridge University Press is part of the University of Cambridge.

It furthers the University's mission by disseminating knowledge in the pursuit of education, learning, and research at the highest international levels of excellence.

www.cambridge.org
Information on this title: www.cambridge.org/9781108845618
DOI: 10.1017/9781108980258

© Anju Mary Paul 2022

This publication is in copyright. Subject to statutory exception and to the provisions of relevant collective licensing agreements, no reproduction of any part may take place without the written permission of Cambridge University Press.

First published 2022

A catalogue record for this publication is available from the British Library.

Library of Congress Cataloging-in-Publication Data
Names: Paul, Anju Mary, author.
Title: Asian scientists on the move : changing science in a changing Asia / Anju Mary Paul.
Description: Cambridge ; New York, NY : Cambridge University Press, 2022. | Includes bibliographical references and index.
Identifiers: LCCN 2021026957 (print) | LCCN 2021026958 (ebook) | ISBN 9781108845618 (hardback) | ISBN 9781108970082 (paperback) | ISBN 9781108980258 (ebook)
Subjects: LCSH: Science – Asia – Case studies. | Research – Asia – Case studies. | Scientists – Education (Higher) – Asia. | Scientists – Education (Higher) – Western countries. | Asians – Education (Higher) – Western countries. | BISAC: SOCIAL SCIENCE / Sociology / General
Classification: LCC Q127.A65 P38 2022 (print) | LCC Q127.A65 (ebook) | DDC 509.5–dc23
LC record available at https://lccn.loc.gov/2021026957
LC ebook record available at https://lccn.loc.gov/2021026958

ISBN 978-1-108-84561-8 Hardback

Cambridge University Press has no responsibility for the persistence or accuracy of URLs for external or third-party internet websites referred to in this publication and does not guarantee that any content on such websites is, or will remain, accurate or appropriate.

To
Appa and Amma,
who returned to India when they did not have to
and
who taught me what a life of service looks like

Contents

List of Figures	*page* viii
List of Tables	ix
Preface	xi
Acknowledgments	xviii
Glossary	xxi
List of Abbreviations	xxiii
Part I Contexts	1
1 Introduction	3
2 Four Case Studies of Science in Asia	41
Part II Circulations	81
3 Leaving Home, Heading West	83
4 Learning Science in the West	115
5 Return to the Future or the Past?	147
6 Asian Women Scientists on the Move	183
Part III Consequences	219
7 New Scientific Research Systems in a Changing Asia	221
8 Shifting Scientific Cultures in a Changing Asia	258
9 Conclusion	287
Bibliography	306
Index	333

Figures

1.1	Two possible career courses in academic science	*page* 13
1.2	The Asian scientist migration system in the second half of the twentieth century	19
1.3	The Asian scientist migration system in the early twenty-first century	20
1.4	The two components of the scientific research environment in a country/organization	39
4.1	Seven dimensions of scientific cultures	121
5.1	Traditional polarity of axes of influence around the return decision for Asian scientists in the West	152
7.1	Five dimensions of scientific research systems	224

Tables

1.1	Twelve most common doctoral and postdoctoral training sites	*page* 28
1.2	Twelve most common places of employment of interviewees who returned	30
1.3	Country of birth of interviewees, by gender	31
1.4	Number of interviewees, by decade of first departure from Asia and first stage of training in the West	32
1.5	Number of returnees, by country of birth	33
3.1	Number of interviewees, by Western country and decade of first arrival	106
4.1	Number of interviewees by Western destination, first training stage in the West, and institution type	116
5.1	Number of returnees, by country of birth and decade of return	156
5.2	Number of returnees, by country of birth and country of interview	158
5.3	Distribution of returnees, by number of children at time of return	166
5.4	Distribution of children of returnees, by age range at time of return	166
5.5	Rank of returnees upon return, by country of return	175
6.1	Descriptive statistics of female interviewees, by country of interview	192

Preface

When nonacademics ask me why I chose to research the migrations of Asian academic scientists, I usually tell them that I had been intrigued by the news stories about the "rise" of Asian science which began appearing with increasing regularity from the early 2000s. The fact that I am a returned academic migrant myself is not something I share, even though my personal experience of return migration played a big part in my decision to study the brain circulations of Asian-born, Western-trained bioscientists.

While I was born in India and am an Indian citizen, I moved to Singapore in 1992 when I was sixteen – after winning a Singapore government scholarship – to complete the last two years of my high school education. My parents supported my pursuit of this scholarship as they recognized that my interests did not fit well with the rather rigid educational tracks that existed for middle-class youth in India at the time, where medicine and engineering were often deemed to be the only acceptable pursuits for academically gifted youth. After I finished high school in Singapore, I won another scholarship to pursue an undergraduate degree in Singapore, but this time it came with a service bond with Singapore Airlines that required me to work for the company for six years after graduation. I was drawn to the idea of funding my own college education without having to ask my parents for financial support, so I accepted the scholarship. After graduating from the National University of Singapore (NUS) with a first-class honors degree in Business Administration, I worked in Singapore Airlines for five years as a management executive.

It took me a while to figure out what I wanted to do with my life and, in the spirit of exploration, I left Singapore in 2003 to pursue a graduate degree in journalism in the USA. I never imagined that I would return to Singapore years later. After my master's, I stayed on in the USA to pursue a doctoral degree in sociology and public policy. My doctoral dissertation focused on the stepwise migrations adopted by

Filipino migrant domestic workers who seek to reach countries higher up their destination hierarchy – often countries in the West – by working in various "stepping stone" countries for several years at a time. When successful, this stepwise migration allows these capital-constrained migrants to accumulate resources and work experience, which they can leverage to climb up their destination ladder.[1] The year I was finishing my dissertation, I heard about a new liberal arts college being set up in Singapore as a partnership between NUS and Yale University. Yale-NUS College was hiring faculty across the humanities, and the social and natural sciences. I applied, hoping that my prior connection to Singapore and my research on Asian migrations would make me a strong candidate. I received an offer to join Yale-NUS as an assistant professor and returned to Singapore in 2013 as a freshly minted academic with my husband and two young children in tow.

Working at a brand new liberal arts college housed within a public research university considered one of the top universities in Asia, gave me the opportunity to witness firsthand the rapid changes underway in tertiary education in Singapore and elsewhere in Asia. While universities in the USA and much of Europe were complaining about declining state support for higher education,[2] many Asian governments were pumping more and more funds into raising the research profile and world rankings of their national universities.[3] They were also actively recruiting Western-trained Asian academics (mainly in the sciences) back to their shores. Seeing these efforts piqued my curiosity and instigated this book project. What made an Asian scientist living in the West uproot themselves (and, in some cases, their families) and move back to Asia? Were there other academics like me – born in one Asian country but "returning" to work in a different one? Was return to a rapidly changing Asia a sign of failure on the part of the returning scientist, or a sign of their ambition? These were the questions I wanted to explore with this project.

But there was another return story that informed my research. This was my parents' return migration to India when I was a teenager.

My parents are both medical doctors – my father is a gynecologist, while my mother was trained as a radiotherapist. Born in the southern

---

[1] Paul 2011, 2017.
[2] Hyatt et al. 2015.
[3] Paul and Long 2016; Knight 2008; Marginson et al. 2013.

Indian state of Kerala to a tenant farmer, my father was the second oldest of nine siblings. He had always been bright in school and, early in his life, he encountered mentors who guided his education trajectory and pointed him towards medicine as a career. After finishing high school, he sat for the Kerala state entrance examinations and was accepted into Kottayam Medical College, a government-run medical school in central Kerala.

My mother came from a more well-to-do family. On her maternal side, her grandparents owned a local bank as well as large swathes of land in the center of the state. Her paternal grandparents traded in spices. Meanwhile, my mother's parents were both English teachers, and believed in the importance of education for all five of their children. My mother, the oldest child, was encouraged to aim as high as possible and she chose to pursue medicine, earning a place at Christian Medical College (CMC) in Vellore in the neighboring state of Tamil Nadu. CMC had been set up by American missionaries during the British colonial period and, in the 1960s, was considered one of the top medical colleges in the country.

My parents met and married after they completed their medical training in the late 1960s. Both joined the Kerala government's medical college system to work in public medical college hospitals, where they practiced medicine and taught the next generation of Kerala's medical doctors. Every few years, they would be rotated to a different medical college hospital in a different part of the state and provided government quarters to live in. During this time, they also gave birth to two daughters – my older sister and myself. But the idea of migrating westward for further training was always a dream at the back of their minds.

By this time, the USA had opened up its immigration system to allow highly educated professionals from Asia (and elsewhere) to enter the country,[4] and there was a mad rush among Indian doctors to apply for jobs in America. By the late 1960s, the Indian government had become so concerned about this "brain drain" that it banned the USA from conducting its medical certification examinations – the Entrance Certificate for Foreign Medical Graduates – within the country.[5] Interested Indian doctors had to travel to neighboring countries in

---

[4] Donato and Amuedo-Dorantes 2020.
[5] Bobb 1977.

order to sit for these qualifying examinations. My father borrowed money from friends to fly to Sri Lanka to take the exam. He passed, received a job offer from a hospital in Texas, and was issued an immigrant visa by the US embassy in Madras (now known as Chennai) in southern India.

But my father was not interested in settling down in the USA. His plan from the very start was to train overseas and then return home. Worried that going to the USA on an immigrant visa would encourage him to settle permanently in the West, he eventually turned down the job offer. My parents ended up working in India for another three years, slowly rising through the ranks within the local medical college system. But my father's dream of pursuing advanced training in the West remained. In 1977, the United States tightened its immigration process for foreign-trained doctors, but the United Kingdom (UK) did not yet require Indian-trained doctors to clear additional certification tests. So my father decided to go to the UK instead.[6]

My father's first job in the UK was a temporary one in a hospital near Belfast in Northern Ireland. He traveled there on his own in 1980 while my mother, my sister and I stayed behind in India. After his Belfast stint, he landed a permanent job in Edinburgh in Scotland and sent for us to join him. We arrived in Scotland in late 1980 and spent three happy years in Edinburgh with my mother choosing to become a stay-at-home wife. My brother was born in Edinburgh, while I picked up a thick Scottish accent. At work, my father rose from resident to senior resident, and was offered a one-year position as a consultant, which was the most senior rank for doctors at his hospital. Such temporary consulting positions often led to a permanent one. Once again, he understood that if he accepted this offer, we would most likely settle down in the UK and not return to India. And so, my father chose to accept a position as head of department in a brand new hospital in the city of Dammam in the Eastern Province of the Kingdom of Saudi Arabia. We left the UK and moved to Saudi Arabia in 1984.

In the early 1980s, Saudi Arabia was still in the midst of its massive development boom as a result of the 1973 oil price hike initiated by the

---

[6] In the 1950s and 1960s, the UK had recruited Indian doctors in large numbers. Though the volume of this migration stream shrank in the 1970s, it was still relatively common and certainly aspirational for Indian doctors to engage in westward migration to the former colonial "metropole" in order to seek further training.

Organization of Petroleum Exporting Countries, and the further price increase after the Iranian Revolution in 1979. Saudi Arabia was building infrastructure at a furious pace and importing large numbers of foreign manpower to build this infrastructure. Lacking a native-born educated professional class, Saudi Arabia was hiring high-skilled foreign labor as well, initially from other countries in the Middle East, but increasingly from South Asia. My father was part of the first wave of educated Indian engineers and doctors who moved to Saudi Arabia and other oil-rich Arab countries in the 1980s, attracted by the high salaries, good benefits, zero income taxes and proximity to India. We stayed in Saudi Arabia for six years and, while I missed my friends in Edinburgh, I enjoyed my sheltered life in Dammam as well. We were embedded within a large and vibrant Indian expatriate community, lived in an apartment building where all of our neighbors were other Indian doctors and their families, and went to an Indian school that followed the Indian curriculum.

After six years in Saudi Arabia, my parents decided that it was time for us to return to India. I did not know this at the time but they had taken a ten-year leave of absence from their state government jobs, and they were reaching the end of their leave period. They needed to return if they still wanted to access their civil service pensions. In addition, they wanted to be closer to their extended family in Kerala as their own parents were getting old. And perhaps most importantly, my father had never given up on his desire to return to work in India where he felt that the need for his skills and training was the greatest.

In early 1990, my mother returned to India, taking us children with her. Meanwhile, my father stayed on his own in Dammam for a few more months to finish out his contract. Saddam Hussein invaded Kuwait soon after we left. I interpreted the subsequent Gulf War as a sign that our family's time in Saudi Arabia should come to an end. After my father returned to India, we settled down in the town of Thrissur in central Kerala, where my father served initially as Associate Professor, and later as Professor and Head of the Department of Obstetrics and Gynecology, at Thrissur Medical College. I had never been interested in medicine as a career, but I would learn from conversations over the dinner table about the challenges and joys he experienced practicing and teaching medicine in India. His government college hospital was always short of funds, so my father learned to innovate new medical procedures and design new surgical instruments under tight budget constraints. As department head in Thrissur, he introduced many of the operating procedures and management practices he had learned in Scotland. Over the years, he trained successive generations of young Indian doctors to approach the practice of obstetrics and gynecology in a more systematic and research-driven manner. He was sometimes frustrated as departmental politics were always present, but it was also an incredibly rewarding experience. He later established the first statewide confidential review of maternal mortality in India, leading to a significant decline in maternal deaths throughout Kerala.

Outside of work, my father would complain about the inefficiencies of the Indian government bureaucracy and the rule-breaking behavior of Indian drivers. A stickler for honesty, he hated the fact that black money was widespread throughout the Kerala economy and that too many people seemed to be on the lookout for a kickback or a handout.[7] My mother meanwhile found it hard to adjust to working-life back in India after ten years of not working and so took early retirement, devoting her energy to managing the household and my father's parallel private practice.

I did not stay long in India. As mentioned earlier, I won a scholarship that took me to Singapore at the age of sixteen. From Singapore, I moved to the USA where I met and married my husband, and had two children, before eventually returning to Singapore. Many of the themes I raise in this book – scientific remittances, gender compromises, scientific cultures, the tension between private ambition and national duty – had parallels in my parents' return migration story as well as my own. The focus of this book – the change in Asia's position within the global scientific field – was only just beginning when my parents returned to India in 1990, but it was in full swing by the time I returned to Singapore in 2013. At its heart, *Asian Scientists on the Move* is about the "things" – aspirations, plans, ideas, values, preferences, perspectives, connections, and knowledge – picked up and carried by circulating migrants and how these social remittances inevitably affect the culture of the communities these

---

[7] "Black money" refers to illegally earned income that is not declared in income tax filings.

migrants return to when they come "home." For that reason, I hope that this book will resonate with the many Asians – like my parents and I – who left their home countries to pursue education and careers in the West, and the increasing number of Asians – again, like my parents and I – who are choosing to come back.

Acknowledgments

This book project has been a long time coming. But none of it would have happened without the generous funding support I received from the Global Asia Institute (GAI) at the National University of Singapore (NUS). Gavin Jones, who headed the GAI at the time I moved to Singapore, was a kind and open-minded supporter of ambitious research projects. He let me take this project in unexpected ways and always ensured that there was additional funding to support my expanding research goals. Likewise, Yale-NUS College generously stepped in to provide a bridging grant when I finally ran out of GAI money before I had finished my fieldwork.

Conducting the fieldwork for this project was so much fun, even as it was exhausting. Over the course of four years, I traveled to India, China, Taiwan and the United States to conduct interviews and visit various research centers associated with the life sciences. In Singapore, I conducted my interviews at NUS, Nanyang Technological University (NTU), and Biopolis, and became intimately familiar with the geographies of their respective campuses. In Taiwan, I spent time at the Academia Sinica campus, visiting various life science institutes, as well as the campuses of National Taiwan University in Taipei and the National Health Research Institutes in Miaoli County. In China, I visited various Chinese Academy of Sciences (CAS) institutes across Beijing. And in India, I conducted face-to-face interviews in Bangalore at the National Centre of Biological Sciences (NCBS), the Indian Institute of Science (IISc) and other research institute campuses. In order to reach further afield in all four of my Asian case countries, I also conducted Skype interviews with scientists in other locations. This was how I conducted most of my US interviews which were scattered throughout the East Coast, the West Coast, and also the Midwest. I continue to be humbled by the generosity of all the scientists I interviewed. They opened their offices and labs to me, they introduced me to their spouses and they told me about their childhood joys and professional challenges. It is because of them that this project took the shape it did, and became much more interesting than I could have ever hoped.

Supporting me while I conducted this fieldwork was a group of stalwart research assistants: Xiao Yun, Li Qin, Pearlyn Neo, Victoria Long, Simonas Bartulis, Nanlan Li and Regina Hong. All were fully committed to the project and each brought their particular strengths and enthusiasms to the endeavor. Later, Anastasiya Varenytsya was invaluable in the copyediting of the final manuscript. Even my children, Sebastian and Paloma, helped with some of the data analysis, while Sebastian read through the entire manuscript and gave me useful feedback.

Among my faculty peers, I have benefited from countless conversations with many interlocuters over the years. At Yale-NUS, I presented chapter drafts to the Gender Research Cluster and the Race, Ethnicity and Migration Cluster, and I am grateful to Huey Shy Chau, Nienke Boer, Gretchen Head, Christine Walker, Cecilia Van Hollen, Gabriele Koch, Kurt Kuehne, Robin Zheng and Zachary Howlett for their insightful comments and suggestions each time I shared my work with them. My life science colleagues at Yale-NUS – Eunice Tan and Ajay Sriram Mathuru – kindly read an early draft of Chapter 6 and gave me helpful feedback. I also presented my work at external venues through the kind invitation of academics whom I respect and admire. Monamie Bhadra Haines, Hallam Stevens and Ian McGonigle invited me to present my work on scientific cultures at NTU in Singapore; Johan Lindquist invited me to give a talk on the trailing wives of returning Asian scientists at the University of Stockholm; Sanna Saksela invited me to the University of Helsinki to present my research on the gender compromises made by Asian women scientists; Hein de Haas, Mathias Czaika and Sorana Toma invited me to a workshop at Oxford University to talk about my initial findings on changing training patterns among aspiring Asian scientists. Other scholars – including Teo You Yenn, Helga Nowotny, Devesh Kapur, Arne Westad, Parvati Raghuram, Peggy Levitt and Yasmin Ortiga – offered advice and suggestions for additional readings and new frameworks to use, and I am grateful to all of them for their support.

Throughout it all, my editor at Cambridge University Press, Joe Ng, was there. He had taken a bet on me with my first book, *Multinational Maids*, and was willing to do it again with *Asian*

Scientists on the Move. Joe – you have been a stalwart presence from beginning to end, always patient with me no matter how many deadlines I missed. Thank you. I must also thank the rest of the team at Cambridge – Chloe Quinn, Felinda Sharmal and Alexander Macleod – for their joint efforts to ready my manuscript for publication.

Finally, I must thank my family. Many times, my husband and children travelled with me when I flew to different countries to conduct interviews. Their memories of visiting farms and museums around the world are interwoven with my memories of interviews in grey offices and cluttered labs. I will be forever grateful for Sebastian and Paloma's understanding when I would tell them that I have to write instead of play with them. My biggest supporter is, of course, my husband Eduardo, who took care of the kids when I could not, and who brainstormed ideas with me for each chapter. I end by thanking my parents to whom this book is dedicated. Their lived example of public service and their decision to return to India has shaped me in ways that I only fully realized as I worked on this book. Thank you for all that you have done and still do for me.

Glossary

Academia Sinica The National Academy of Taiwan, headquartered in Taipei

Gaokao (Chinese term) National College Entrance Examination in China

Haigui (Chinese term) A colloquial term used in China to describe Chinese emigrants who study overseas for several years and then return to China. Similar in pronunciation to the Chinese word for "sea turtle," which travels for long distances but always returns home

Hundred Talents Program A program established by the Chinese Academy of Sciences (CAS) to recruit established, as well as young, overseas Chinese academics to return to work at one of the CAS institutes in China

Impact-Factor (Also known as journal impact factor) Calculated from the yearly average number of citations garnered by articles published by the journal in the previous two years, it is often used as a measure of a journal's importance within a particular field

Ivy League A group of eight, elite and exclusive, private research universities in the northeastern USA. They include Harvard University, Princeton University, Yale University, Columbia University, Cornell University and others

Nature A British peer-reviewed academic journal that is considered one of the top scientific journals in the world

One China Policy The Chinese policy that there is only one China as opposed to two – the People's Republic of China and the Republic of China (Taiwan)

Project 211 A Chinese program launched in 1995 to improve the quality of education and research at approximately 100 universities throughout the country

Project 985 A Chinese program launched in 1998 to transform some of China's top universities into "world-class" research universities

Science A peer-reviewed academic journal of the American Association for the Advancement of Science, it is considered one of the top scientific journals in the world

Thousand Talents Program Also known as the Thousand Talents Plan, this program was established by the Chinese government in 2008 to attract overseas researchers to work in China on either permanent or short-term appointment contracts

Abbreviations

A*STAR	Agency for Science, Technology and Research (Singapore)
ASEAN	Association of Southeast Asian Nations
ATREE	Ashoka Trust for Research on Ecology and the Environment (India)
BBS	Bulletin Board System
BCE	Before the Common Era
CAS	Chinese Academy of Sciences (China)
CCMB	Centre for Cellular and Molecular Biology (India)
CCP	Communist Party of China
CMC	Christian Medical College (India)
CSIR	Council of Scientific and Industrial Research (CSIR)
CUSBEA	Chinese-US Biochemistry Examination and Application
DBT	Department of Biotechnology (India)
EU	European Union
FDI	Foreign Direct Investment
GAI	Global Asia Institute (Singapore)
GDP	Gross Domestic Product
GPA	Grade Point Average
GSK	GlaxoSmithKline
IBS	IndiaBioscience
IISc	Indian Institute of Science
IISER	Indian Institute of Science Education and Research
IIT	Indian Institute of Technology
IMCB	Institute of Molecular and Cellular Biology (Singapore)
INR	Indian Rupees
IP	Intellectual Property
KMT	Kuomintang Party
MBBS	Bachelor of Medicine, Bachelor of Science (undergraduate medical degree common in the UK and Commonwealth countries)

MD	Doctor of Medicine (graduate medical degree common in the USA)
MD+PHD	Doctor of Medicine and Doctor of Philosophy
MIT	Massachusetts Institute of Technology (USA)
MOST	Ministry of Science & Technology (Taiwan)
NBRP	National Biotechnology Research Park (Taiwan)
NCBS	National Centre for Biological Sciences (India)
NHRI	National Health Research Institutes (Taiwan)
NIH	National Institutes of Health (USA)
NSF	National Science Foundation (USA)
NSTB	National Science and Technology Board (NSTB) (Singapore)
NT$/NTD	New Taiwan Dollars
NTU	Nanyang Technological University (Singapore)
NUS	National University of Singapore (Singapore)
OECD	Organization for Economic Co-operation and Development
OPT	Optional Practical Training (USA)
PhD	Doctor of Philosophy
PI	Principal Investigator
PRC	People's Republic of China
R&D	Research and Development
RMB	Renminbi (the official currency of China)
ROC	Republic of China (the official name of Taiwan)
SBS	School of Biological Sciences (Singapore)
SGD	Singapore Dollars
STEM	Science, Technology, Engineering and Mathematics
STS	Science and Technology Studies
THE	Times Higher Education
TIFR	Tata Institute of Fundamental Research (India)
TIGP	Taiwan International Graduate Program
UK	United Kingdom
UNESCO	United Nations Educational, Scientific and Cultural Organization
US/USA	United States of America
USD/US$	US Dollars

Note: when it is not clear where a particular abbreviation originates, the relevant country is included in parenthesis.

PART I
Contexts

1 Introduction

The rising volume of scientific research coming out of Asia has been making headlines since the start of the twenty-first century.[1] Much of this attention has focused on the rapid ascent of China, with plentiful news stories and reports about the amount of funding the Chinese government has invested in upgrading its scientific research system.[2] In 2007, China overtook the United States (USA) to become the largest producer of natural science and engineering doctorates in the world. In 2015, China produced 32,000 doctorates in these fields, while the USA produced 30,000.[3] China's overall research and development (R&D) expenditure now exceeds that of the European Union (EU) as a whole.[4]

Several other Asian countries have also been expanding their investments in scientific research and education in the last two decades and they have paid particular attention to the life sciences. In 2003, Singapore opened its multibillion dollar Biopolis campus to serve as the base for several of the country's public research institutes that focus on the biomedical sciences. The Asian R&D offices of several multinational pharmaceutical businesses were also set up on the Biopolis campus.[5] In 2008, the South Korean government launched its 577 Program to boost the proportion of the country's Gross Domestic Product (GDP) spent on R&D to 5 percent, and promised to funnel

---

[1] Woetzel and Seong 2020; Veugelers 2012; *Nature* 2007. There is even an *Asian Scientist* magazine (www.asianscientist.com) that publishes articles highlighting research and researchers from Asia. It was launched in 2011. The journal *Asia Pacific Biotech News*, started in 1997 by World Scientific Publishing, focuses exclusively on life science research in the Asia Pacific region (www.asiabiotech.com/about-us.html).
[2] Xie, Zhang and Lai 2014; Veugelers 2017; Zhou 2015; Zhang, Sun and Bao 2017.
[3] Khan, Beethika, Carol Robbins and Abigail Okrent, "The State of U.S. Science and Engineering 2020," National Science Board, January 15, 2020, figure 11, accessed on January 12, 2021, https://ncses.nsf.gov/pubs/nsb20201/global-r-d.
[4] Khan et al., "The State of U.S. Science and Engineering 2020," figure 11.
[5] Chan 2006; Cyranoski 2001; Smaglik 2003.

these funds equally to basic and applied research in seven key technology areas, including the life sciences.[6] In 2018, Taiwan opened its National Biotechnology Research Park (NBRP) adjacent to Academia Sinica, the country's premier base for scientific research.[7] Together with the Hsinchu Biomedical Science Park, which is about an hour's drive south of Taipei, the NBRP is expected to raise Taiwan's competitiveness in biotechnology. In 2010, India began setting up a series of new universities along the lines of its renowned Indian Institutes of Technology (IIT), but now with a focus on science education and research.[8] Called the Indian Institutes of Science Education and Research (IISERs), these seven universities offer students a joint bachelor's and master's degree in various scientific fields – including the life sciences. These are just some of the many centrally led initiatives launched by different Asian governments over the last two decades – all aiming to propel a rise in their respective country's global scientific standing. As a result of these efforts, Asia's overall share of the world's R&D investment is now larger than that of the Americas and Europe, and it continues to grow.[9] More and more patents are being filed in Asia and the number of scientific journal articles published by scientists based in Asian countries has also increased considerably.[10]

But behind this science story, there is a *migration* story. The elite scientists who are leading these new research institutes and carrying out this cutting-edge research in Asia are rarely "homegrown" talent.[11] Instead, the vast majority of the scientific personnel fueling the current boom in Asia's scientific research output were born in Asia but trained in the West, primarily in the USA but also in the United Kingdom (UK), continental Europe and Canada. Only after the completion of their training, and sometimes after several more years of working as academic scientists in the West, did they return to work at the top research universities and institutes in various Asian countries. Vanya, an Indian

---

[6] Stone 2008.
[7] *Taiwan Today* 2018.
[8] Stone 2012; *Nature* 2016.
[9] Heney 2020; Grueber and Studt 2013.
[10] Grueber and Studt 2013; King 2004; Leydesdorff and Zhou 2005; Veugelers 2017.
[11] Paul and Long 2017; Cerna and Chou 2014; Ortiga et al. 2018; Yeoh and Lai 2008. Israel has adopted a similar return policy, encouraging highly skilled emigrants or its "knowledge diaspora"(Welch and Hao 2016) who are believed to be able to contribute to the national economy to return (Cohen 2009).

scientist, is one such Western-trained returnee. She now works at a top research institute in India where all of her colleagues have a similar migration history. As she put it, "My entire research institute is full of people who have all moved back [to India]. So, you know, I am very much par for the course. Every single one of them has been abroad and then come back."

Like me, Vanya had a parental history of international migration and return that had influenced her own migration decisions.[12] When they were young, Vanya's parents left India to train in the West. Their decision to return to India in the 1970s was considered unusual by their peers and relatives. Vanya shared the difference in people's reactions to her parents' return decision in the 1970s versus her own return in the 2000s:

> When they came back [to India] in the 1970s, hardly that many people were coming back. Now, that's not the case. Now, I'm not unusual at all.
>
> I remember my parents being asked, "Why are you coming back? How foolish are you? You know, if you're medical doctors, if you live in the USA, you will make a ton of money and live a very smooth and simple life."
>
> But now, nobody asks us that much. I mean, people do ask it, but it's a lot less given that there are so many people who make that move [back to India]. It's not unusual now.

In the present day, most returning Asian scientists go back to work in their country of birth, but other returnees are choosing to work in another Asian country. Singapore, in particular, has been a beneficiary of what I call "halfway-return" migration that brings elite Asian-born, Western-trained scientists back to Asia, but to an Asian country other than their birth country.[13]

Why did these Asian scientists choose to train in the West? Has the logic around training locations changed in recent years? What made

---

[12] I wrote about my own multigenerational history of international migration in the Preface.

[13] Even though "halfway-return" is not quite grammatically correct, I use this term as this is what one of my interviewees used to describe his journey from China to the USA for doctoral and postdoctoral training, and then from the USA to Singapore to start work in a public research institute. "Halfway-return" speaks to a third option that is situated somewhere in-between staying in the West versus a "true" return to one's birth country.

some of these scientists choose to return to Asia after their training, while others remained in the West? Is the calculus around this return decision changing? How does their gender and nationality affect Asian scientists' experiences in the West and their return logics? What happens after return, when Western-trained Asian scientists set up labs in Asia? What do these scientists bring back with them when they return? These are the questions motivating this book. To answer them, I draw on in-depth interviews I conducted with 119 elite Asian bioscientists who trained at top universities in the West, 86 of whom had returned to Asia – though not always to their country of birth. These 119 bioscientists are drawn from a range of Asian countries but the vast majority come from my four main Asian fieldsites – China, India, Singapore and Taiwan.

From my interviews, I identified four interlinked developments in the Asian "corner" of the contemporary global scientific field. The first development relates to recent upgrades of the scientific research systems in select Asian countries. These improvements in research systems are driven in a top-down manner by the governments of these countries investing significant public funds into scientific R&D. As a result of these improvements, the volume of return migration to Asia is increasing as ambitious Asian scientists no longer see Western countries as the only viable base from which to launch a successful research career.

But returning scientists require graduate students and postdoctoral fellows – the foot soldiers of science – to staff their labs, and so they (and their governments) are now seeking to expand the availability and improve the robustness of local scientific training options, rather than outsource the task of training to Western countries. This leads to the second finding of this book which is the increasing diversification of training pathways emerging within the Asian scientist migration system. The consequence of macro- and micro-level efforts to improve the scientific training infrastructure in select Asian countries is that more aspiring Asian scientists are delaying when they leave home for training in the West, with increasing numbers leaving only at the postdoctoral moment, rather than earlier at the doctoral training moment. Intra-Asian mobility is also increasing for training, networking and career progression purposes. And so, while more aspiring Asian scientists may choose to stay and train in their own country of birth, others are instead moving to another Asian country that offers attractive packages for trainees and is closer to home than the West. Western countries still play an important role in these trainees' professional lives, but not in the same way that they used to twenty years ago. This book traces all of these dynamic changes in the Asian scientist migration system.

The third change I uncover is how returned Asian scientists are attempting to affect the scientific research systems and scientific cultures in the top Asian research organizations where they return to work. By tracing elite Asian scientists' journeys to the West, and exploring their experiences there as well as back in Asia (for those who chose to return), this book identifies the "scientific remittances" these returning scientists bring back with them. Having reached a critical mass, returning Asian scientists are now driving change from the ground up in their Asian research universities and institutes. I investigate the changes these elite scientists are seeking in how labs are run, teams managed, science is taught and applied, and how the next generation of Asian scientists is being mentored. I show that while Asian governments may have been focused on making structural changes to their domestic scientific research systems – from improving the quality of available research technology, to increasing the funding for scientific research – they instigated a more fundamental change in scientific cultures, resulting from the new values and perspectives that returning Asian scientists are bringing back with them.

My fourth finding is the extent of variation that exists *across* the scientific research systems and scientific cultures in Asia, and the differing challenges that my four Asian case countries face as they seek to enhance their relative standing in the global scientific field. These challenges operate at the level of scientific research systems, but also in terms of the specific scientific cultures that exist in individual research organizations. These are described firsthand by the scientists I interviewed in my four Asian case countries. As a result, I am able to differentiate between the scientific terrains in each of my case countries, rather than treat Asia as a monolith. Similarly, this book does not treat the West, or the practice of science in Western countries, as a uniform whole. Given that my interviewees trained in a range of Western countries, their interviews highlight the differences between various Western countries' scientific research systems and scientific cultures, debunking the idea of a single "Western science," in addition to the idea of a single "Asian science."

My hope is that, through all this, *Asian Scientists on the Move* will give readers an insight into the rapidly changing global scientific field – and particularly the Asian corner of this field – without allowing readers to fall prey to simplistic East–West narratives. From a theoretical point of view, this book is distinct from migration scholarship that sometimes sidelines developmental and other large-scale societal processes that shape and are shaped by migration.[14] In contrast, this book embeds its analysis of the migrations of individual academic bioscientists from different parts of Asia within the larger story of their particular country's scientific development, showing the mutually influencing relationship that exists between development and migration even among the highly skilled.

This book also contributes to science and technology studies (STS) with its detailed descriptions of the scientific training environments in my four Asian case countries and how these environments have changed since the 1980s and 1990s. This is possible thanks to the composition of my interviewee sample which includes successive generations of elite Asian scientists who left their home country to go to a Western country for training in three different time periods (the 1980s and earlier, the 1990s, and the 2000s), and successive cohorts of returnees as well. But before I can introduce my interviewees, I need to first situate the object of my study: the global scientific field, and Asia's place within it.

A Social Field Analysis of the Global Scientific Field

This book explores the global scientific field: how it has changed in recent years, how scientists from Asia physically and figuratively navigate this field, and how four Asian countries – China, India, Singapore and Taiwan – are working to change their relative position within this field. In order to do all this, I start by explaining what a social field is, the social actors and norms that comprise the global scientific field, what holds symbolic value within this field, and how the field interacts with other social fields (in particular, the "field of power")[15] to shape scientists' career course and life course.

---

[14] De Haas 2010.
[15] For this section, I draw primarily on Pierre Bourdieu's work on field analysis. It was Bourdieu who insisted that a key part of field analysis is understanding how

Concepts for Analyzing the Global Scientific Field

French sociologist Pierre Bourdieu defined social fields as segments of the broader social space that possess their own particular social structure, their own internal logic determining how status is achieved within the field, and some degree of autonomy from the broader society.[16] Social fields are comprised of social actors (or agents) who could be individuals or formal/informal groups, organizations, or even larger entities such as states. These agents hold different positions within the field depending upon their possession of, or access to, the particular resources that have been deemed worthy within the field. These resources, or "capitals" as Bourdieu termed them, can take different forms and the list of possible capitals has only increased and gained more specificity since the time Bourdieu first outlined three key capitals – economic, social and cultural.[17] Bourdieu also emphasized that capitals have symbolic value only in relation to particular fields. In the context of the scientific field, "scientific capital" has been identified as a science-specific version of these three and other capitals. Scientific capital includes:

(1) Research funds (economic capital) to buy scientific equipment and materials, and hire research staff,
(2) Advanced training (human capital) in scientific research methods, tools and know-how,
(3) An ease and familiarity with the history, norms and values of one's subfield, and institutional affiliation with key organizations in the subfield (cultural capital), and finally,
(4) Network contacts with and recognition by the field's gatekeepers and other researchers in one's subdiscipline (social capital).[18]

The possession of one or more of the above types of scientific capital imbues an individual with status within the global scientific field. But scientific capital can also be used to acquire and accumulate *additional* status within the field, primarily through publishing in high impact-factor journals. This is a version of the Matthew Effect, first

---

any given social field relates to the broader "field of power" (Bourdieu 2005:33; Bourdieu and Wacquant 1992:97).

[16] Bourdieu 1993; Archer et al. 2015.
[17] Bourdieu 1986.
[18] Bourdieu 1988; Archer et al. 2015.

coined by sociologist Robert Merton (1968), which posits that in science, as in life, those with more get ever more, while those with little get less.[19] In the biological sciences, the top-ranking journals to which ambitious academic scientists submit their articles include *Nature* (and its sister journals, including *Nature Cell Biology* and *Nature Genetics*), *Science*, *The Lancet*, *Cell*, the *Annual Review of Plant Biology*, *Genome Biology*, and *Trends in Ecology and Evolution* – to name a few. Other ways to garner respect and recognition in the field are successfully securing large research grants, and winning prizes and filing patents stemming from one's discoveries. Additional capital-linked criteria used to determine status within the scientific field include one's rank, the reputation of the research organization where one works, the overall size of one's research support team and research laboratory, the number of one's citations, and involvement in high-profile research collaborations.

Even as scientists regularly collaborate on joint research projects, Bourdieu wrote that every social field is a "field of struggle," where agents are constantly jockeying for greater relative positions within the field through the accumulation, exchange and monopolization of field resources.[20] However, even as they are in competition with each other, all agents in the field are united by a sense of the importance of their struggle (or the "game" as Bourdieu termed it). All the actors also have an unspoken agreement over how this game should be played and what should count as status markers within the game.[21] This collective acceptance of the rules of the game is critical because it is what effectively makes the field a field and demarcates its boundaries. At the same time, however, each field is dynamically changing with either new actors attempting to enter the field, or existing actors working to

[19] Merton 1968.
[20] Bourdieu 1993. Another form of "struggle" that can occur within a scientific field is a clash of paradigms to explain particular scientific phenomena. Thomas Kuhn's (1962) *Structure of Scientific Revolutions* is perhaps the most well-known account of such a struggle within the scientific field when a long-standing paradigm comes under increasing threat as more and more scientists within the field begin to accept a new way of looking at particular phenomena/evidence. The discipline may enter a period of crisis as this clash of paradigms continues, until the new paradigm replaces the old one in what Kuhn called a "scientific revolution." I do not deal with paradigmatic conflicts in *Asian Scientists on the Move*; instead, I study the much more traditional struggle for power and status within the scientific field.
[21] Lenoir 2006.

raise their relative position within the field by accumulating new capital or changing the rules of the game to block new entrants.

Within the global scientific field, the actors involved are myriad and operate at different levels. They include states and their relevant government ministries,[22] public and private universities and research institutes, corporations that house a scientific research arm to develop new technologies and products or fund academic research collaborations, and also individual scientists and students training to become scientists. Within the contemporary scientific field, particular countries (almost all in the West) are viewed as occupying the center or "core" of the field, meaning that most of the key institutions and individuals within the field are based in these countries.[23] After World War II, the USA in particular grew in prominence within the global scientific field as the single largest producer of science and engineering doctorates, and the largest individual funder of scientific R&D in the world. Asian countries were largely situated on the periphery or semi-periphery of the global scientific field for most of the twentieth century, resulting in a gravitational force that encouraged the westward migration of aspiring Asian scientists. But, as I show in this book, the topography of the global scientific field is shifting as select Asian countries are investing significant economic resources – acquired through their strong performance in other fields or by diverting funds from other sectors – into improving their standing as producers of scientific research.

Within the global scientific field, I focus on the life sciences which cover the various branches of science concerned with the study of life and living organisms. The life sciences encompass a broad range of subdisciplines from ecology to genetics and everything in-between. The life sciences can be considered a field in and of itself. Likewise, the scientific actors, institutions and research systems in each country can also be considered to constitute their own geographically contained field, situated within the overarching global scientific field.

---

[22] Typically, this would include a ministry/department of education, and a ministry/department of science and technology, though this unit is sometimes subsumed under a ministry of trade and economy.
[23] Dear 2001; Jacob 1997; Jasanoff 2005; Xie and Killewald 2012.

Intersecting Temporalities

When studying the particular spatiality of the global scientific field, it is necessary to engage with the unique temporality of academic science. In *An Orderly Mess*, former president of the European Research Council Helga Nowotny notes that "the time generators of societies reside in [social] institutions. Their temporal regimes coordinate, regulate and impose the temporal grids and timelines that structure our lives" (2017:20). Academic science has a specific temporality and requires a particularly demanding investment of time and resources from aspiring academic scientists. They are expected to undergo a lengthy period of study and training before they can embark upon a career as an academic scientist. There used to be a time, more than a hundred years ago, when a scientist could be self-taught.[24] But in the present day, an academic scientist needs to prove their worth with formal qualifications garnered through years of study and training at recognized tertiary educational institutions.

At the same time, the typical "career course" for academic scientists varies from country to country and does not always follow a fixed or stable progression.[25] It is often presumed that an aspiring scientist would study a particular scientific discipline for their undergraduate degree, then pursue further graduate training and eventually secure a doctoral degree in the same field (see Figure 1.1). For scientists who seek a career in academia at the top research institutions in their field, postdoctoral training lasting anywhere from one to six years is also required.[26]

The biological sciences have led the explosion in the number of postdoctoral trainees in the USA over the last few decades.[27] Within the biological sciences, there were 6,866 postdoctoral appointees in the USA in 1979; by 2015, this number had increased threefold to 21,261

---

[24] Jacob 1997; Thurs 2007.
[25] This variation was noted as far back as 1918 when Max Weber (1958) lectured about "Science as a Vocation."
[26] National Research Council 2014, 2000; Cantwell 2011; Alexander von Humboldt Foundation 2013. The US National Science Foundation (NSF) (2014) defines a postdoctoral appointee as "an individual who has received a doctoral degree (or equivalent) and is engaged in a temporary and defined period of mentored advanced training to enhance the professional skills and research independence needed to pursue his or her chosen career path" (National Research Council 2014).
[27] Alexander von Humboldt Foundation 2013; Auriol 2010; Stephan et al. 2013.

A Social Field Analysis of the Global Scientific Field 13

Possible career course leading to a position at a research university:

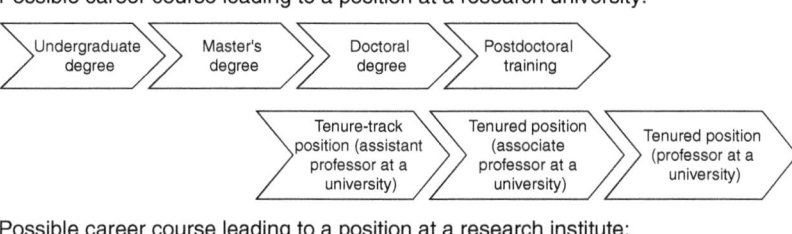

Possible career course leading to a position at a research institute:

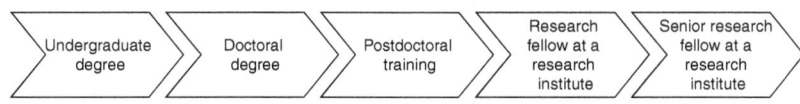

Figure 1.1 Two possible career courses in academic science

(including those in neuroscience).[28] This increase was driven by the growing belief in the biological sciences that postdoctoral appointees are essential for the efficient and productive running of a scientist's laboratory. As research projects become more complex, team-based and expensive, postdoctoral trainees provide a much-needed buffer for their scientist-supervisors, allowing the scientist or principal investigator (PI) to remove themselves from many day-to-day supervisory and training tasks.[29]

Supply-side factors have also contributed to the mandatory nature of postdoctoral training in the biological sciences in the West. Until the 1970s, a postdoctoral appointment was not required for a doctorate holder in the USA to secure a tenure-track faculty position in the sciences.[30] From 1970 to 1974, only half of doctoral degree recipients in the biological sciences reported plans for postdoctoral study.[31] However, the 1980s saw an explosion in the number of doctoral degrees (PhDs) being produced each year in the USA, without

---

[28] "Survey of Graduate Students and Postdoctorates in Science and Engineering, Fall 2015," National Science Foundation, table 28, accessed on June 20, 2021, https://ncsesdata.nsf.gov/datatables/gradpostdoc/2015/html/GSS2015_DST_28.html.
[29] Cantwell 2011; Moguérou 2005.
[30] Zumeta 1985; Cantwell 2011.
[31] "Survey of Earned Doctorates and Doctorate Records File," National Science Foundation, figure 6-11, accessed on June 20, 2021, https://wayback.archive-it.org/5902/20150817174927/http://www.nsf.gov/statistics/nsf06319/.

a concomitant increase in the number of faculty positions available.[32] This resulted in a hiring bottleneck that increased the number of PhD holders taking up a temporary postdoctoral appointment in order to publish additional articles and strengthen their résumé before applying for a faculty position.[33] With the increasing specialization that is also occurring in the biological sciences, a postdoctoral training stint is today viewed as essential for a career as an independent researcher in academia in all top research universities in the West and Asia,[34] while the duration of these postdoctoral training appointments has also lengthened significantly.[35]

Upon completion of their training, US-based scientists who want to join academia may seek to join a university or college as an assistant professor on a multiyear contract, then apply for tenure and promotion to associate professor after several years, and eventually seek a further promotion to the rank of full professor (see Figure 1.1). As faculty, these scientists are expected to teach and mentor undergraduate and graduate students in their laboratories and classes, in addition to pursuing their particular research agenda and publishing their findings in scientific journals. Other aspiring academic scientists may choose to take up a position at a public/private research institute, which imposes fewer teaching obligations and allows them to focus most of their time on research (see Figure 1.1).

In both cases, these scientists would set up a research laboratory and form a research team comprising short-term trainees (who could be students or postdoctoral fellows) and, in some cases, more permanent staff (such as a lab manager). While they would receive some start-up and annual funding from their institution, they would often still need to seek additional funds from various funding agencies at the regional, national or even international levels to take on sufficient staff and purchase the necessary equipment and supplies to pursue their particular research agenda. These research grant applications require a significant amount of a scientist's time to prepare and then manage (if successful). As such, a scientist's research team is vitally important in enabling the scientist to conduct experiments, collect and analyze data, write articles showcasing

---

[32] National Research Council (NRC) 2000.
[33] National Research Council 2014.
[34] National Research Council 2014, 2000.
[35] Regets 1998; Cantwell 2011; Moguérou 2005.

the results of their studies, and otherwise maintain their research productivity.

The career courses described in Figure 1.1 are only two of many possible professional trajectories for academic scientists. The order, duration and valence of each stage will vary across countries. In the USA, for instance, many scientists report progressing directly from an undergraduate degree to a doctoral program that folds a master's education within it. In Europe and parts of Asia, in contrast, students often pursue a standalone master's degree after completing an undergraduate degree, and only then apply for shorter PhD programs which include limited coursework.

In certain Chinese research organizations, returning scientists who have completed at least four years of postdoctoral training overseas are able to take up the rank of full professor directly, without having to go through the earlier stages of assistant and associate professor. Meanwhile, most universities in the USA have the social institution of tenure which effectively grants an academic job security for life after they clear a performance review – usually at the six-year mark. The pressure to "publish or perish" is very high on US-based assistant professors as they work towards their tenure review. This is especially true at the top research universities in the country where publishing expectations keep ratcheting up. Likewise, in Singapore's top two universities – the National University of Singapore (NUS) and Nanyang Technological University (NTU) – tenure standards are also increasing, resulting in a significant amount of stress surrounding the tenure moment for junior academics in the country. As in the USA, assistant professors in Singapore who fail to receive tenure are required to leave their institution the following year. In India, on the other hand, there is little stress about job security as it is generally expected that once you have been hired by a public university, you will have the job until retirement.

But it is not only that a scientist's career progression is spatially variant; the relative status attached to each of these stages also depends upon the country. A full professorship at an American university carries higher symbolic value within the global scientific field compared to a full professorship at an Indian regional university, even if the titles and ranks are technically the same. These spatially determined variations in the status and stress attached to each career stage can also have implications for the patterns of brain circulation that develop within this field, as I explore later in the book.

In addition to failing to receive tenure, there are other reasons why Figure 1.1's steady, automatic and unilinear progression from one career stage to another is not guaranteed. In some cases, individuals may "fall" into a scientific career after pursuing other subjects or other careers (such as medicine). In other cases, scientists may find themselves stuck in a particular career stage and unable to progress to the next because of a lack of research productivity, organizational discrimination, or a host of other reasons. A scientist's career course also intersects with and is impacted by other temporal scales, in particular their life course. The life course is a temporal scale embedded in social institutions – such as puberty, marriage and parenthood – which often have a biological element to them, but are also imbued with socially generated ideas about what are normative behaviors and priorities at each stage in the life course. The intersection of their career course with their life course has a significant impact on women scientists, often forcing them to make "gender compromises" that may have a negative impact on their career progression.[36] Nowotny refers to "cracks in the timelines" (2017:21) that individuals experience when different temporal regimes they are embedded in and progressing through clash with one another. The word "crack" implies a stoppage, a break, an inability to progress further along one's timeline. For Asian women scientists, I find that this can be an all-too-frequent occurrence when their life course crashes into their career course. However, women scientists are not the only ones who experience cracks in their timelines. In this book, I also explore the much rarer gender compromises that male scientists occasionally make and the conditions under which they make such compromises.

Scientists may also experience "jumps" in their timelines, which I define as the acceleration of a particular stage in their career course. This could also occur because of an intersection between their career course and life course, or it could occur because of a migration. A central question of this book is what migration to the West, and also migration back to Asia, does to an aspiring scientist's career course. Does it speed up their career progression? Are they able to "jump" time on their career course by "jumping" physical space? Or does migration result in cracks in their career progression through the global scientific field? And what do all these

---

[36] Ackers 2004; Monosson 2011.

high-skilled migrations do for the science and technology trajectory of scientists' origin and destination countries?

Brain Circulations through Space and Time

In the 1960s and 1970s, there was rising concern in developing countries that their "best and brightest" were decamping for the West, and particularly the USA, rather than staying in their birth country and helping to build their own nation's institutions and economy.[37] In the scientific realm, "brain drain" from the Global South and the Communist Bloc was viewed as leading to "brain gain" in select Western countries.[38] While some of these migrating scientists were fleeing political persecution or overbearing state control, many others were simply looking for a more comfortable lifestyle and greater resources to pursue their particular research agenda. Aspiring scientists from countries in Asia and Africa would often first leave on student visas to train in the West but then stay on to work in Western research organizations. These westward scientist migrations were seen by origin countries as a zero-sum game until the 1990s, when a new term "brain circulation" came into vogue in migration studies as well as in policy circles.[39]

Tied to ideas about transnational migration, the premise behind brain circulation is that the out-migration of highly skilled individuals should not be treated as a net loss for their origin country because these individuals often remain deeply connected with their country of birth.[40] These highly skilled migrants may start transnational businesses that employ people from their birth country, donate to charitable causes in the home country or, through their successful integration

---

[37] Docquier and Rapoport 2012; Bhagwati and Hamada 1974; Kapur and McHale 2005. The term "brain drain" was first coined to describe the migration of scientists and other high-skilled individuals from the UK to the USA in the 1950s and 1960s. The fear of brain drain from Europe to the USA has not gone away, as evidenced by the 2021 upset over the failure of the Pasteur Institute in France and the French pharmaceutical giant Sanofi to develop a successful vaccine against Covid-19 when Moderna, the US-based pharmaceutical company that did successfully develop a vaccine, is headed by a French national, Stéphane Bancel (Willsher 2021).

[38] Ioannidis 2004; Hunter et al. 2009; Ganguli 2014.

[39] Saxenian 2002, 2005; Le 2008; Robertson 2006; Velema 2012; Ackers 2005.

[40] Basch et al. 1994; Glick Schiller et al. 1992, 1995; Ackers 2005.

overseas, help create a positive impression of the birth country in their adoptive country. AnnaLee Saxenian used the term to describe the ways in which Taiwanese and Indian engineers who went to the USA for graduate study, subsequently joined Silicon Valley companies and then used their position in these companies to build transnational outsourcing links with companies in their birth countries. Some of these emigrants eventually returned to their birth countries to manage branch offices of their US companies, or to set up new businesses that leveraged their American ties to drum up new transnational business. In this manner, these transnational migrants helped funnel a significant volume of investment dollars from the USA to Taiwan, India, and more recently, China. While such high-skilled emigrants had previously been looked at as almost traitors to the homeland for having settled overseas, they began to be viewed more positively by policymakers in their origin country. The recognition that emigrants could contribute economically, not just through financial remittances but in more substantive ways, and that sometimes they could even be encouraged to return home many years later led to active outreach by Asian governments. Countries would send overseas missions to the popular destinations of their high-skilled emigrants to cultivate social, cultural and business ties with these individuals, and encourage them to return or to invest in the origin country.[41]

In the scientific realm, the brain circulation pattern connecting East and West in the second half of the twentieth century involved an uneven but still somewhat mutually beneficial scientist migration system (see Figure 1.2).[42] Large numbers of aspiring scientists from Asia travelled to Western countries for training because of the lack of sufficient training infrastructures in their home country. Most of these trainees chose to eventually settle down in the West and, in particular, the USA, where they helped propel scientific output from the West to ever greater heights. A small number of these trainees returned home to staff universities and research universities in Asia, where they encouraged their own students to go to the West for advanced training,

---

[41] Tzeng 2006.
[42] I intentionally use the terminology of a "migration system" (Bakewell 2014; Mabogunje 1970) to highlight the relationships connecting multiple locations in the world-system, and to allow for a multi-level analysis of the operations of this system which can be relatively stable but still incorporates feedback loops and external inputs that allow for change within the system over time.

A Social Field Analysis of the Global Scientific Field 19

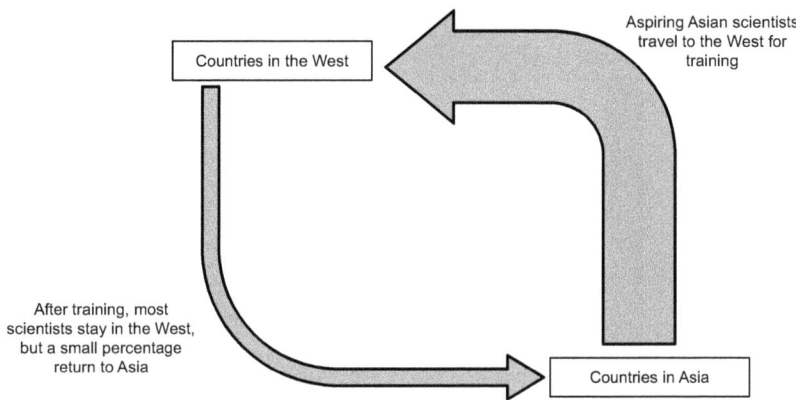

Figure 1.2 The Asian scientist migration system in the second half of the twentieth century

restarting the cycle of westward scientist migration once again. Some Asian governments would even provide scholarships for their top undergraduate students to pursue training in the West, under the condition that these students would return to work in the home country upon the completion of their graduate studies.

However, the Asian scientist migration system is undergoing change in the twenty-first century (see Figure 1.3). As select Asian countries invest significant funds into upgrading their scientific research systems and their national universities, the viability of advanced scientific training in these Asian countries is increasing, as are the prospects for conducting high-impact scientific research in Asia itself. These developments are diversifying the directionality and temporality of Asian scientist flows. A key contribution of *Asian Scientists on the Move* is uncovering these new brain circulation patterns emerging in the global scientific field. These new developments include delays in the moment of first departure to the West for scientific training, the halfway-return of Western-trained Asian scientists to an Asian country other than their country of birth, as well as the increase in intra-Asian scientific mobility.

Scientific Remittances and Scientific Cultures

With the growing numbers of Asian scientists returning to work in Asia, it is worthwhile asking what these scientists bring with them when they

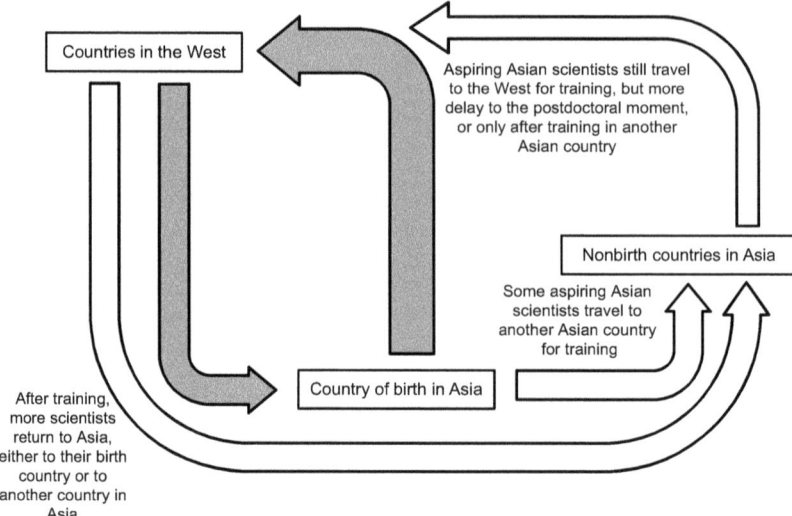

Figure 1.3 The Asian scientist migration system in the early twenty-first century

come back to Asia. In this book, I develop the idea of "scientific remittances" to capture all that overseas scientists send back to the scientific communities in their origin country, and also what returning scientists bring with them when they move back home.[43] Scholars of high-skilled migration have shown how emigrant scientists facilitate the development of transnational research collaborations between scientific institutions in their destination and origin countries. The same can be said for scientists who return after spending several years overseas where they established research connections with other scientists. Scientific remittances encompass such transnational network ties that scientists carry with them, but these remittances also reference the circulation of new scientific know-how, reputational standing, and norms and values regarding the social practice of scientific research.

It is this last component of scientific remittances – new scientific norms and values – that I dwell on the most in this book. Even as scientists themselves might emphasize the universality of the scientific method and position themselves as sitting somewhat outside of and

---

[43] Here I build on the idea of social remittances pioneered by Peggy Levitt (Levitt 1998; Levitt and Lamba-Nieves 2011).

independent from society, STS scholars have long observed that scientific research is, in fact, a social practice that is fully embedded within and influenced by society.[44] Along these lines, Robert Merton made a useful distinction between the "methods of science" and the "mores of science." He defined the latter as the "prescriptions, proscriptions, preferences, and permissions" that are "legitimized in terms of institutional values" (1973 [1942]:269) within scientific circles. As a sociologist of science, he was especially interested in the specific institutional norms that are internalized by scientists to form what he called the "scientific conscience." The four norms he identified were:

(1) Universalism (which contends that scientific claims must be judged objectively without any consideration for the gender, race, class, nationality or any other identity dimension of the claimant, and likewise, that scientific careers should be open to all on the basis of their capabilities and competence – and nothing else),
(2) Communality[45] (in the sense that the knowledge produced by scientific research should be shared rather than hoarded),
(3) Disinterestedness (based on the expectation that scientists pursue knowledge without any consideration for personal or other types of gain), and
(4) Organized skepticism (which expects a critical standpoint toward any truth claims until they have been objectively verified).

In reality, scientists regularly deviate from Merton's four scientific norms. For instance, there is widespread acknowledgement that personal ambition plays a pivotal role in scientific research, with widespread competition to be the first to publish a field-changing paper or make a new discovery being played out at multiple levels – between individual scientists, the universities and institutes where they work, and even the countries they belong to.[46] Likewise, the argument that one should pursue "science for science's sake" is no longer left unquestioned by

---

[44] Merton 1973 (1942); Nowotny 2017; Latour 1987; Kuhn 1962; Reckwitz 2002.
[45] Merton actually termed this norm "communism," but later scholars rephrased it as "communality" which carries less political baggage (Barber 1962; Storer 1966).
[46] Merton 1973 (1968). James D. Watson's (1968) *The Double Helix* is perhaps the most well-known firsthand account of a scientist's desire for glory and fame within the field in the midst of pursuing knowledge for all.

funding agencies and even by members of the public.[47] The principle of universalism in science is also routinely violated, as scientists have been shown to be just as human as everyone else in society, carrying implicit and explicit biases against members of various social groups (whether along gender, race, ethnicity, caste or nationality lines, to name just a few of the bases of discrimination that occur within scientific circles).[48] Even the principle of communality has been tested at times, particularly when national interests intervene in areas of scientific research deemed to have security implications or commercial possibilities. This has led to concerns about naturalized immigrant scientists sending sensitive information about new technologies back to their birth countries.[49] Rather than treating Merton's four scientific norms as inviolable, it makes more sense to think of these characteristics of scientists and scientific teams as variables that operate on a spectrum.

As scientific research becomes increasingly team based, it is important to consider additional social and cultural norms and values that affect interpersonal dynamics and individual performance within a given scientific group.[50] Organizational cultures shape how people in a group or team engage in work, and whether or not their full capabilities and potential are realized. Londa Schiebinger writes that scientific cultures encompass "rituals of day-to-day conformity, codes governing language, styles of interactions, modes of dress, hierarchies of values and practices" (1999:68). Together with the scientific research system in which it is embedded, the scientific culture of a lab or an organization plays a critical role in an aspiring scientist's personal growth and professional success. A central focus of this book is trying to understand the specific cultural norms that returning Asian scientists bring back with them that are related to how they think research teams should be formed, labs managed, and students of science mentored and taught. For this reason, I disaggregate scientific cultures along seven key dimensions: their attitude towards scientific knowledge, their approach to problem-solving, the scope of their research ambitions, the degree of autonomy that individual scientists in the research team are given and expected to have, the level of importance given to rank and seniority within the scientific environment, attitudes towards difference within the group, and finally, the communication norms of the

---

[47] Weber 1958; Nowotny et al. 2001; Gibbons et al. 1994.
[48] Loder 1999; Thomas 2020; Subramaniam 2015.
[49] Wagner 2021.
[50] Franklin 1995; Traweek 1992, 1993; Krautwurst 2014.

research team. I explore the different scientific norms that Asian scientists are exposed to during their overseas training, their adoption of some of these norms, and the ones they bring back with them when they return to start their own labs in Asia. In this manner, *Asian Scientists on the Move* deals with the cultural diffusion that is a consequence of Asian scientists' return migration. Large-scale cultural change takes generations and is only just beginning in the select Asian organizations I studied. But what my interviews show is that this change is underway and it is being driven by the return migrations of Western-trained Asian scientists, as much as by broad-based societal change in Asian countries.

Research Design and Methods

This project started in 2013, when I was awarded a three-year grant from the Global Asia Institute at NUS to study the return migration decision-making processes of Asian scientists from India, China, Singapore and Taiwan. All four of these Asian countries (and others, such as Japan and South Korea) are competing in a "global war for talent" to lure scientists in the Science, Technology, Engineering and Mathematics (STEM) fields to their shores.[51] I chose those four countries to provide two paired comparison frames, the first between India and China, and the second between Taiwan and Singapore.

Case Selection

India and China were chosen because they are the two largest countries in Asia, with roughly equivalent populations and comparable scientific innovations in ancient times, followed by a long period of scientific stagnation for much of the early and mid-twentieth century. If anything, India might have been considered slightly ahead of China at its moment of independence and for a few decades afterwards. The tragic failures of the Great Leap Forward and the Cultural Revolution kept China mired in poverty until the late 1970s, when Deng Xiaoping took over leadership of the Chinese Communist Party and ushered in a series of sweeping economic and educational reforms. During the same period, India's leaders pursued a policy of import substitution and strong protectionism against free trade and foreign investment, stifling

---

[51] Oishi 2012; Murakami 2014; Lewin and Zhong 2013; Cerna and Czaika 2021.

its economic growth. It was only in the early 1990s that India began to liberalize its economy, leading to a rapid growth in its middle-class population. Fast forward to the twenty-first century, and it is clear that China has moved well ahead of India in terms of economic growth as well as investment in scientific research, though India has picked up its pace in recent years. While both of these Asian giants have encouraged the return of native-born scientists trained in Western universities, return migrations to China have been significantly larger in volume, lured by a better funded and more ambitious government effort. In choosing these two countries, I wanted to understand why returning to China seemed easier for Chinese scientists compared to Indian scientists' return to India, and if there were lessons in China's return policies and recent upgrades in its scientific research programs that could be applied in India.

My second frame of comparison was between Taiwan and Singapore, two relatively small island-nations in Asia. Taiwan has a population of twenty-three million while Singapore is even smaller with just over five million. Both countries were labelled Asian tiger economies in the 1980s, after experiencing rapid economic development jumpstarted through manufacturing-linked foreign direct investment (FDI) by multi-national corporations from the USA, Europe and Japan. Both are considered developmental states with strong governmental involvement in macroeconomic planning. In the 1990s, both also experienced a rapid rise in their cost of living, leading to their respective governments attempting to shift the national economy towards more knowledge-driven industries. Taiwan had an initial advantage over Singapore, as many Taiwanese pursued graduate education in science and engineering in the USA, with a significant proportion voluntarily returning to Taiwan as its economy took off. Singapore, being smaller and lacking a large diaspora of native-born scientists, could not rely as much on its own populace to boost its scientific output. Nonetheless, in the first decade of the twenty-first century, Singapore was being fêted as a rising global powerhouse in bioscience research, while similar hype was not used to describe Taiwan. The differential success of these two smaller Asian countries intrigued me, and I wanted to explore what factors had facilitated Singapore's rapid rise in the global scientific field and whether or not the return migration of Singaporean scientists had played any role.[52]

---

[52] South Korea was the other East Asian country that was deemed an Asian tiger economy, and it too pivoted strongly towards biotechnology at the same time as

Definitions and Scope Conditions

In studying the brain circulations of Asian scientists, I focus on scientists in the life sciences. By limiting my analysis to a subpopulation of scientists rather than all STEM scientists, I was attempting to study a smaller and more uniform group of high-skilled returnees.[53] This allowed me to conduct a more coherent and in-depth analysis of the field-specific factors behind my interviewees' migration decisions and postreturn experiences. This approach also allowed me to focus on specific research organizations within each country case study, without stretching myself too thin. And the reason I chose to focus on the life sciences was because several governments in Asia and elsewhere had dubbed the twenty-first century as the "century of biology,"[54] investing significant funds to boost their domestic bioscience research output. That said, the life sciences encompass a wide range of subdisciplines, including plant biology, evolutionary biology, molecular biology, biostatistics, ecology and bioengineering, to name just a few.

Within the life sciences, I limited my analysis to *academic* scientists who were working in public/private research universities or research institutes. Doctoral recipients who subsequently pursued a research career in "industry," such as in a pharmaceutical company or some other for-profit corporation, were not the subjects of interest for my project, as they have different expectations about their professional trajectory and different industry-led research goals. In order to qualify for inclusion in this project, an interviewee had to hold a PhD, engage in academic research in the life sciences or related fields in either a public/private university or research institute, and be permitted to serve as the PI on a research grant application. This last criterion effectively excluded postdoctoral trainees, who typically work under the supervision of a faculty member who serves as the PI on their grant applications. I did interview a handful of Asian postdoctoral trainees over the course of this project, but they are not counted in my sample of 119.

Taiwan and Singapore. However, with a population of more than fifty-one million people, it does not have quite the same challenges as Taiwan and Singapore in developing its domestic biotech industry.

[53] Chompalov 2006; Ackers 2005.
[54] Biotechnology and Biological Sciences Research Council 2013; National Research Council 2009; Wong 2005, 2011.

When I discuss "scientists-in-training," I am referring to doctoral students and postdoctoral fellows in the life sciences. The inclusion of postdoctoral fellows in the category of scientists-in-training recognizes that, within the life sciences, a multiyear postdoctoral fellowship is now considered *de rigueur* before an individual can apply for an independent position as a faculty member or PI in a research university and institute.

I use the term "Asian scientist" to refer to an individual who was born in East, South or Southeast Asia. I do not count anyone from Southwest Asia (which is more commonly known as the Middle East), Russia or other members of the Commonwealth of Independent States in this category. While my primary recruitment sites in Asia were China, India, Singapore and Taiwan, I also encountered scientists who originally came from other parts of Asia (including Japan, South Korea, Pakistan and Malaysia) who were working in one of my four Asian case countries or in the USA. (These scientists are counted in my sample of 119 but they only constitute 11 interviewees out of the whole sample.)[55] When I talk about the "Western" training received by the Asian scientists I interviewed, I am referring to training that occurred in the USA, Canada or any country in Western Europe. Australia, New Zealand, Israel and former Soviet bloc countries are left out of this definition of the West. Throughout this book, both "Asia" and the "West" are used in a purely geographical and geopolitical sense, and do not imply any kind of cultural uniformity across these large, socially constructed territorial categories. I stress this point because of the danger that this book could be mistakenly linked to ethnocentric claims about the superiority of Western science over Asian science. I do not subscribe to the view that one can talk about either the West or Asia as having a single scientific culture, and I discuss this in more depth in Chapter 4 and Chapter 8.

Research Methods

My primary research method involved in-depth, one-on-one interviews with Asian-born, Western-trained bioscientists. Participants were recruited in four Asian countries – China, India, Singapore and Taiwan – and also in

[55] I also interviewed 65 Western-born and Western-trained scientists who moved to work in Asia, but they are not the focus of this book, even though their West to East migrations are another emergent brain circulation pattern that is emblematic of the changes underway in the global scientific field.

the USA between 2014 and 2016. Potential participants were identified by visiting the public websites of top bioscience research institutes and of bioscience departments in top research universities in the five country fieldsites. These websites typically list all the faculty affiliated with the institute/department, along with their educational and employment histories. Scientists who possessed names that are typically associated with nationality/ethnic groups from Asia, or had an undergraduate degree from an Asian university, followed by a PhD and/or a postdoctoral fellowship in a Western university, were sent an email introducing the study and participation criteria, and asking if they would be willing to participate in a one-hour confidential interview. As a token of appreciation for participating in the study, scientists were offered a US$50 Amazon.com voucher.

Whenever possible, scientists' spouses were also invited to be interviewed so that I could learn more about the family decision-making process around the question of return. In several cases, these spouses were also bioscientists and were then included in my primary sample if they were also working in academia.

All interviews were conducted in English. The interviews were semi-structured and all of them covered the following topics: the interviewee's family background and educational history in their native country, their decision to leave for training in the West, their experiences in the West, their decision to return to Asia (or not), their experiences back in Asia (for those who chose to return), and their future plans. Interviewees' responses to each of these questions are used to structure this book, with each chapter focused on one of these questions in chronological order. Chapter 6 is the one exception. That chapter focuses on the experiences of Asian women scientists navigating the global scientific field. It explores how these women's gender identity intersects with other identity dimensions they may hold (such as their nationality, ethnicity and religion). I chose to have a separate chapter on Asian women scientists because I wanted to call readers' attention to the ways in which their gender completely redefines Asian women's pursuit of a career in science.

Interviewee Profiles

Before offering some descriptive statistics about my interviewees, I want to reemphasize that, as a group, they would be classified as elite scientists, on the basis of the doctoral and postdoctoral

Table 1.1 *Twelve most common doctoral and postdoctoral training sites*

University	Country	Number of interviewees	
		With PhDs from the site	With postdoctoral training from the site
Harvard University	USA	3	8
Cornell University	USA	6	3
Stanford University	USA	3	5
Yale University	USA	3	4
Cambridge University	UK	3	3
Kyoto University	Japan	3	2
University of Illinois at Urbana-Champaign	USA	3	2
Indian Institute of Science	India	3	1
University of Pennsylvania	USA	1	3
Massachusetts Institute of Technology	USA	0	4
Columbia University	USA	0	3
University of California, Berkeley	USA	0	3
Total		28 (24%)	41 (39%)[*]

[*] Not all interviewees received postdoctoral training. This percentage is derived from the number of interviewees who completed at least one postdoctoral training stint.
Note: Training sites are ordered according to the total number of interviewees who completed either a PhD or a postdoctoral training stint at the site, in declining order. Only training sites where three or more interviewees received either their doctoral or postdoctoral training are included in this table.

institutions where they trained, as well as the scientific institutions where they were working at the time of their interview with me. Table 1.1 lists the twelve most common doctoral and postdoctoral training sites of my interviewees to give an indicative sense of the research organizations where they trained. In the USA, this list included universities like Harvard, Cornell, Stanford and Yale – all private universities and all ranked in the top twenty universities in the world, according to the 2021 *Times Higher Education* (*THE*) World

University Rankings.[56] In the UK, it included universities like Cambridge and Oxford – which are ranked in the top ten universities in the world. Fewer interviewees received their doctoral training in Asia but, among those who did, certain universities were more likely to be mentioned. These included Kyoto University in Japan (considered one of the top STEM universities in the country), the Indian Institute of Science (considered the top science university in the country), and Zhejiang University in China (one of the oldest universities in China and a top research university). No interviewee born in Singapore or Taiwan completed their doctoral training in their birth country. The twelve universities in Table 1.1 represent 24 percent of all my interviewees' doctoral training sites and 39 percent of their postdoctoral training sites (for those with postdoctoral training).

A second way to demonstrate the elite status of my interviewees is to consider the research organizations where they were working at the time I interviewed them. Table 1.2 shows the top twelve places of employment of those interviewees who returned to Asia.[57] I mention this so it is clear that when I talk about the changes underway in Asian research universities and institutes, I am referencing the very top institutions in my four Asian case countries. In the cases of China and India – my two large-country case studies – and to some extent Taiwan, these changes may not be indicative of what is happening at more local or regional institutions in these countries. At the same time, even though their experiences are not representative of all scientists in their country, these interviewees do represent the vanguard of change in their respective country's scientific community.

Even as I sought interviews with individuals who constitute the elite stratum of scientists within each of my four Asian case countries, I chose not to interview institute directors and university leaders in these countries. This was because I wanted to understand the on-the-ground, daily experience of conducting scientific research in Asia during the second decade of the twenty-first century, hearing both the

---

[56] "Best Universities in the World," *Times Higher Education* World University Rankings, September 2, 2020, accessed on January 28, 2021, www.timeshighereducation.com/student/best-universities/best-universities-world.

[57] In the case of interviewees who chose not to return to Asia, the most common institutions where they were working in the USA were Columbia University, Scripps Research Institute, the University of Pennsylvania, and Johns Hopkins University.

Table 1.2 *Twelve most common places of employment of interviewees who returned*

University/Institute	Country	Number of returnees
Chinese Academy of Sciences	China	13
Nanyang Technological University	Singapore	12
Academia Sinica	Taiwan	9
National Taiwan University	Taiwan	8
National University of Singapore	Singapore	8
Genome Institute of Singapore	Singapore	5
Indian Institute of Science	India	5
Institute of Molecular and Cell Biology	Singapore	4
National Health Research Institutes	Taiwan	4
National Centre of Biological Sciences	India	3
Ashoka Trust for Research in Ecology and the Environment	India	3
Total		74 (86%*)

* This percentage is derived from the number of interviewees who returned to work in Asia.

advantages as well as the challenges involved. I was concerned that if I spoke with individuals in senior leadership positions, they would be more likely to present a one-sided and overly positive take on the present situation in their country or organization. Thus, I chose to talk with people who were in the so-called trenches, to learn directly from them about the joys and difficulties of conducting scientific research in the life sciences in Asia in the present day, to hear about how conditions have changed since their earlier training years, and to understand the additional changes they would like to see in the scientific research systems and scientific cultures in their respective Asian country.

Table 1.3 provides statistics about the birth countries of the Asian scientists included in this study. As I mentioned earlier, in addition to scientists born in my four main Asian case countries, I interviewed Asian scientists who had been born in other parts of Asia – such as Japan, South Korea, Malaysia and Pakistan – but who were now

Table 1.3 *Country of birth of interviewees, by gender*

Country of birth	Male	Female	Total
China	28	9	37
India	27	9	36
Taiwan	13	11	24
Singapore	4	7	11
Other Asian countries	7	4	11
Total	79	40	119

working in one of my four Asian fieldsites or in the USA. They are classified together under "Other Asian Countries" because of their small numbers.

Table 1.3 also provides a gender breakdown of my sample. While I am happy with the interviews I was able to conduct with Asian women scientists, I acknowledge that I interviewed only half as many women as men. Still, I worked hard to ensure sufficient representation of Asian women scientists in my sample because I wanted to understand how their gender influences their initial migration decision, as well as their desire to return. I particularly wanted to talk with Asian women scientists about their comparative experiences working in the West versus in one or more Asian countries. I focus on their stories in Chapter 6.

Table 1.4 demonstrates the temporal richness of my interviewee sample, which spans Asian scientists who first left their home countries in the 1980s and earlier, to those who left in the mid-2000s. While I did not actively try to recruit interviewees according to their decade of departure, it just so happened that they self-sorted somewhat evenly. As a result, I am able to trace changes in the relative importance of different factors in driving the westward migrations of aspiring Asian scientists across different periods of time. While only 7 percent of interviewees who left Asia in the 1980s and earlier moved at the postdoctoral moment, 30 percent of those who left in the 2000s were moving at the postdoctoral moment. Likewise, while 25 percent of interviewees who left Asia in the 1980s and earlier first left to pursue master's training in the West, by the 2000s only 6 percent of

Table 1.4 Number of interviewees, by decade of first departure from Asia and first stage of training in the West

Decade of first departure	Number of interviewees (%)	Departures by stage of training			
		Undergraduate (%)	Masters (%)	PhD (%)	Postdoctoral (%)
1980s & earlier	28 (24%)	2 (7%)	7 (25%)	17 (61%)	2 (7%)
1990s	44 (37%)	4 (9%)	4 (9%)	28 (64%)	8 (18%)
2000s	47 (39%)	1 (2%)	3 (6%)	29 (62%)	14 (30%)
Total	119	7 (6%)	14 (12%)	74 (62%)	24 (20%)

Note: these numbers represent all 119 Asian-born scientists I interviewed, some of whom returned to Asia, while others chose to stay on in the West.

interviewees were leaving at that stage in their training. These changes speak to the rising standards in science education in various Asian countries vis-à-vis the West, as well as the impact this has on Asian countries' relative position within the global scientific field, and on the training destination decisions of aspiring Asian scientists in the present day and into the future.

Of my 119 interviewees, 86 had returned to work in universities and research institutes in my four Asian country fieldsites (see Table 1.5). However, given that I used a purposive recruitment strategy to ensure that I talked with as many returnees as possible, my sample population's return rate cannot be directly generalized to the universe of Asian scientists. In actuality, the ten-year stay rate of foreign life science and health science doctorate recipients in the USA in 2013 was 67 percent.[58] This statistic reflects the percentage of foreign students who had received a PhD in the life sciences or health sciences in the USA between 2003 and 2004, and who were still working in the USA ten years later in 2013. If we break down the stay rate by origin country, the 2013 stay rates increase to 86 percent for doctorate recipients from China and India.[59] I share this statistic so that readers do not come away from reading this book imagining that most Asian scientists in the

Table 1.5 *Number of returnees, by country of birth*

Country of birth	Total number of interviewees	Number returned to Asia (% of total)	Number returned to birth country (% of returnees)
China	37	27 (73%)	12 (44%)
India	36	28 (78%)	14 (50%)
Taiwan	24	20 (83%)	20 (100%)
Singapore	11	9 (82%)	6 (67%)
Total	108	84 (78%)	52 (62%)

Note: The 11 interviewees born in Japan, South Korea, Malaysia and Pakistan are not included in this table because their numbers were too low. Most of these participants were interviewed in the USA, but two were halfway-returnees whom I interviewed in Singapore.

[58] Finn and Pennington 2018:3.
[59] Finn and Pennington 2018:5.

West return to Asia. This is not currently the case. But what this book does argue is that, compared to the Asian scientist migration system in the second half of the twentieth century, the return flow to Asia has increased in volume and diversified to the point that it is having a significant impact on the research systems and scientific cultures of the Asian research organizations these scientists return to.

Among the returnees I interviewed, not all of them returned to their country of birth (see Table 1.5). While all 20 of the Taiwanese returnees I interviewed had gone back to Taiwan, roughly half of the Indian and Chinese returnees I interviewed were working in another Asian country at the time of their interview. Many of these scientists had opted to work in Singapore as an alternative return strategy for themselves and their families. One such interviewee, a US-trained Chinese scientist who decided to "return" to Singapore rather than China called this move a halfway-return, and, as mentioned earlier, I have adopted his term to describe this hybrid of return and onward migration. Chapter 5 focuses on the emergence of this and other new types of return migration patterns, showcasing how Asia's scientific research terrain is shifting and diversifying in dynamic ways. More statistical data about my interview sample are provided in the following chapters as I dig into specific aspects of their migrations.

Throughout the book, however, I take great pains to keep the identity of my interviewees confidential. I do not indicate at which particular institution they trained or currently work, nor do I offer details about their research specialization when I quote them. All the personal names I use are fictitious. In certain cases, I also do not mention the specific Western country where my interviewees trained if their migration history is so unique as to be identifiable. In other cases, I refer to the same individual by different fictitious names in different chapters so that an astute reader cannot piece together the interviewee's identity from the various pieces of background information I provide throughout the book. While this takes away from a full understanding of each interviewee's story, these precautions are necessary given how small the community of bioscientists are in some of my country case studies. However, I try to compensate for the lack of background information about each interviewee by maintaining their "voice" in their interview excerpts. I engaged in only light editing to improve the clarity of what interviewees told me, while retaining the unique cadence and vocabulary of each of my interlocuters. The excerpts are also often intentionally long because I want my interviewees' ideas and experiences to shine through on the page. My conversations with these scientists were frank, expansive and rich in information. I hope readers come away from this book feeling like they had been in the room with me, listening to these scientists recount their migration experiences.

Structure of the Book

Asian Scientists on the Move is divided into three parts, with Part I ("Contexts") offering the reader an overview of the motivation for and argument of the book, and the key concepts that will be touched upon throughout the book. This first part of the book also provides an overview of the history of and recent changes in the scientific research systems in select Asian countries. In particular, Chapter 2 – "Four Case Studies of Science in Asia" – offers the reader an overview of the efforts by China, India, Singapore and Taiwan to modernize their scientific research sectors, upgrade their national universities into "world-class" research-focused institutions, and woo their native-born scientists to return home. I pay particular attention to how the respective governments of these four countries have attempted to transform their bioscience and biotechnology sectors, and the successes and challenges they have encountered along the way. This overview of the state of scientific research in each of these four countries is a useful counterpoint to Part II ("Circulations"), which offers an on-the-ground perspective on many of these same developments as told from the point of view of individual Asian bioscientists.

The first three chapters in "Circulations" are organized around three key phases in my interviewees' brain circulations and scientific careers. I use Chapter 3 – "Leaving Home, Heading West" – to lay out the multiple, overlapping motivations behind my interviewees' initial migrations to the West. I disaggregate their migration decision into the three constituent subdecisions they needed to make, and explore each one in depth:

(1) *Whether or not* to leave and seek training in the West,
(2) *When* in their scientific training trajectories they should leave, and
(3) *Where* in the West to go for this training.

I also show how overdetermined the desire to seek specialized training in the West used to be. However, because of the unique nature of my interviewee sample, which spans Asian scientists who first left their home countries in the 1980s or earlier all the way up to those who left in the mid-2000s, this chapter also traces how the mix of motivating factors behind a westward migration has begun to shift over time. A key takeaway of this chapter is the growth in the variety of education and training pathways now available to aspiring Asian scientists, such that the dominant pattern of migration in the past – undergraduate studies in Asia followed by doctoral and postdoctoral training in the West – is no longer the only possible (or even most likely) option for an ambitious aspiring scientist from Asia. Finally, I show how the growing numbers of returning Asian scientists is creating internal pressures on current and future cohorts of Asian scientist-trainees to stay in Asia for their training, so as to help staff the research teams of highly ambitious returned scientists.

While Chapter 3 explained why my interviewees left Asia for the West, Chapter 4 – "Learning Science in the West" – describes my interviewees' experiences in the West. I discuss how they found themselves exposed to new scientific cultures – novel ways of thinking about and *doing* science in the Western universities and research institutes where they trained. In this chapter, I offer a conceptual framework for disaggregating any given scientific culture in terms of seven key dimensions: its attitude towards learning scientific knowledge, its approach to problem solving, the scope of its research ambitions, the degree of autonomy that individual scientists in the culture were given and expected to have, the level of importance given to rank and seniority within the research environment, attitudes towards difference in the team, and finally the communication focus of the research team. Going through each dimension in turn, I describe the cultural differences my scientist interviewees encountered at the various institutions and countries where they trained. However, rather than making sweeping statements about Western scientific cultures vis-à-vis Asian scientific cultures, I drill down into variations that operated at the level of individual labs, universities, as well as countries in these two world regions, even as I note general patterns of difference between various parts of the world. My interviewees' exposure to different scientific cultures during their training years fundamentally changed their own approach to scientific research once they became independent researchers in their own right. This transformation influenced their return decisions and (for those who did choose to return) their preferred approach to "doing" science back in Asia.

Out of the 119 Western-trained Asian scientists I interviewed, 52 (or 44 percent) had returned to work in their birth country, while another 34 (29 percent) were working in an Asian country other than their country of birth at the time of their interview. (The remaining 33 interviewees had chosen not to return to Asia.) This tripartite distribution of return options (no return vs. return home vs. return to another Asian country) allows me to dig into the factors that influenced these varying decisions. This is the focus of Chapter 5 – "Return to the Future or the Past?" – which reveals how the process of return was much more complicated for my interviewees than their initial out-migration to the West. I organize this chapter according to the three "axes of influence" that shaped interviewees' return decision – their social and cultural integration into their host society in the West, their professional ambitions and their sense of familial obligation. As in Chapter 3, this chapter interrogates questions about the timing of return and how well it complemented the life- and career-stage of my interviewees, and those of their spouses and children (if they had any). In addition, there was the question of *where* interviewees should return: their birth country or some other Asian country? Their hometown or some other city within their birth country? Linked to this last point, I use this chapter to highlight some of the alternative arrangements that interviewees adopted, which tended to serve as a compromise between the return-versus-no-return binary. These alternatives included leaving the West and moving to an Asian country other than their birth country (a halfway-return), as well as establishing a transnational split-household arrangement where one spouse returned while the other stayed behind in the West. Once again, these arrangements highlight the growing complexity of migration and mobility options being considered by Asian scientists.

The last chapter in this section of the book focuses on the experiences of Asian *women* scientists and the impact of their gender on their scientific training and careers in Asia and the West. Are there particular challenges that Asian women scientists experience in their careers that impact their decisions about where and how to practice science? In order to answer this question, I use this chapter to consider the same set of social institutions and mechanisms I explored in the earlier three chapters but now adopt an explicitly gendered lens to trace how gender affects the international migrations and academic careers of Asian women in science. The forty women scientists I interviewed present a nuanced picture of the current state of affairs in Asia that belie the simplistic stereotype about the region as a whole being biased against women in science. I introduce the new concept of "gender shock," which I coined to describe the experience of entering a social space or milieu that has a set of gender norms and values different from the one that you came from and are familiar with. I highlight how these Asian women scientists were much more likely to experience gender shocks at particular inflection points in their intersecting career and life courses, and how they tended to react to negative gender shocks and ongoing gendered social pressures by making "gender compromises" in one or more domains of their lives. These corrective actions often (but not always) dampened their career trajectories, or took their life course in an unexpected direction. These gender compromises also often involved particular migration and destination decisions.

Throughout these four empirical chapters, I engage with how particular features of the global scientific field create gravitational pressures that propel the westward aspirations and migrations of Asian bioscientists-in-training, but also how these pressures have begun shifting in recent years, leading to changes in the brain circulation patterns between Asia and the West.

In the final section of the book (Part III – "Consequences"), I describe returned Asian scientists' experiences of "doing" science in Asia, and the challenges and surprises they encountered while working in Asian countries that had changed considerably while they were away. This final part of the book focuses on the two components that make up the scientific research environment in any given country/organization: the scientific research system (or structure), and the scientific culture of the place (see Figure 1.4). Chapter 7 focuses on the scientific research systems in my four Asian case countries, while Chapter 8 focuses on their scientific cultures. Both chapters discuss returnees' experiences with these twin aspects of the research environment in their "new" country, and the joys as well as difficulties of conducting research in such settings. Both chapters also document where returnees would like to see further changes to these structures and cultures.

Structure of the Book

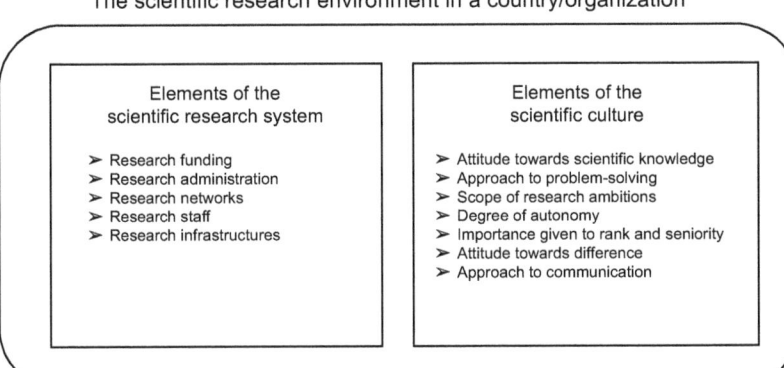

Figure 1.4 The two components of the scientific research environment in a country/organization

I organize Chapter 7 – "New Scientific Research Systems in a Changing Asia" – according to the five key dimensions of the scientific research systems that returnees had to engage with when they began working in Asia (see Figure 1.4). These were the research funding, research administration, research networks, research staff and research infrastructures available to them to support their own research agenda. I discuss where returnees found differences – both positive and negative – between the research systems they had been exposed to while training in the West and the research systems they now encountered back in Asia. I also use this chapter to compare how much the scientific research systems in Asian universities and research institutes have changed over the decades, drawing on the retrospective comparisons made by returnees. From those returnees who went to work in a different Asian country other than their birth country, while still maintaining research ties with their birth country, I am also able to make *intra-Asian* comparisons of different research systems. Through these multipronged comparisons, I explore how variations in the scientific research systems in different Asian countries affect returnees' self-reported research productivity and sense of career fulfilment, and the areas they identified as needing even more change in the future.

In the last empirical chapter of the book – "Shifting Scientific Cultures in a Changing Asia" – I introduce the scientific remittances that returning Asian scientists bring back with them. These include not just new scientific know-how, reputational standing, and scientific network connections, but also new norms and values regarding the social practice of scientific research. While the previous chapter focused on the differences between and changes in research *systems* that returning scientists encountered in Asia, Chapter 8 focuses on how these scientists were trying to effect change in the scientific *cultures* of their new places of employment in Asia. But I also highlight how Asian societies had independently changed, often in significant ways, while my interviewees pursued training in the West. Accordingly, I describe the challenges returned scientists encountered as they tried to change their students' approach to scientific learning, expand their students' research ambitions, level attitudes towards seniority and rank within their research teams, and broaden attitudes towards difference. I show that, even though Asian governments may have been focused on the extrinsic goals of an increase in citations, patent filings and technological leapfrogging when they instigated reverse scientist migration flows, they catalyzed a deeper change. The new values, attitudes and behaviors that returned scientists are trying to inculcate in their students and trainees can potentially lead to unplanned cultural shifts in the Asian research institutes and universities returnees join. I argue that, in the long run, these cultural innovations are just as important as the rise in these organizations' rankings stemming from the return migrations of Asian scientists.

I end the book with a concluding chapter that outlines the theoretical implications of my findings and revisits the new concepts and ideas I introduced throughout the book, which have extended relevance for the fields of migration studies, STS and also gender studies. I highlight what is yet to be studied on this topic, and lay out a future research agenda for scholars from these fields. I also highlight the policy implications for Asian and non-Asian countries from these developments, and offer a set of policy recommendations for government officials and research leaders in these countries as they seek to make themselves attractive destinations for native (and nonnative) research scientists and raise their national profile in the global scientific field.

2 | *Four Case Studies of Science in Asia*

My four Asian field sites – China, India, Singapore and Taiwan – vary significantly in terms of their national history, social structure and political setup. And yet, they share three key similarities. In recent decades, all four have engaged in initiatives to revamp their higher education systems and make their national universities more research focused. All four have also chosen to "bet on biotech,"[1] investing significant public funds to boost their national research production in the biological sciences. Finally, all four are attempting to lure their Western-trained scientists to return "home" with the promise of large research grants and other forms of research support.

This chapter takes a macro-level perspective and explores how the central government in each of these countries approaches the goal of rapid scientific progress in the life sciences. This chapter also discusses the unique hurdles each government faces in trying to achieve this goal. To aid readers with limited background knowledge of Asia, each country case study starts with a brief overview of the history and politics of the country, before discussing its higher education sector, its scientist migration system, its science and technology sector (with a particular focus on bioscience research), and the present and future challenges the country is facing as it seeks to boost its bioscience research output.

This chapter thus provides a brief social field analysis of the biological sciences in each of my four country cases, and also of how this field interacts with other overlapping social fields (such as the higher education sector) within the country and also at the global level. A significant amount of scholarship already exists on these four countries, as well as on each of the topics I am writing about within each country. Given the space constraints of this chapter, I will inevitably not include the finer details about each of these topics, but the

---

[1] Wong 2011.

references in the footnotes will enable curious readers to find additional sources for a deeper study. There are fewer *comparative* studies of different Asian countries' approaches to scientific development however.[2] By placing these four country case studies side-by-side, this chapter helps readers make direct comparisons across these countries to see parallels and contrasts in their approach to achieving largely similar national goals.

China

The People's Republic of China (PRC) is the largest country in Asia in terms of landmass and the largest in the world in terms of population size. The majority of its people are ethnic Han, but there are many minority ethnic groups in the country, especially in its western provinces. For thousands of years, China was ruled by a series of dynasties, though they occupied a smaller territory than the country does now. Imperial rule lasted until 1912 when a revolution ended the Qing dynasty and replaced it with the Republic of China under the nationalist Kuomintang Party (KMT). But the KMT had ongoing struggles with the Communist Party of China (commonly known as the CCP), leading to many years of internal conflict. These hostilities were set aside during World War II when China was battling the invading Japanese. After Japan surrendered to the Allies in 1945, the Chinese Civil War between the KMT and the CCP restarted. The conflict ended in 1949 with a decisive CCP victory over the KMT, leading to the communist takeover of mainland China.

The 1949 establishment of the PRC under CCP Chairman Mao Zedong was followed by the implementation of large-scale social programs to transform the country economically and socially. From 1958 to 1962, Mao's Great Leap Forward sought to restructure China's agrarian economy into a communist one based on industry. However, the campaign was an unmitigated disaster leading to the deaths of tens of millions of Chinese, mostly from famine and

---

[2] Joseph Wong's (2011) *Betting on Biotech* is a notable exception. Wong compares the biotech programs of Singapore, South Korea and Taiwan, the different developmental strategies they adopted, and the challenges they faced in achieving their goals. The role played by high-skilled migration is given less emphasis in his book, though it is still mentioned. I benefited greatly from his analysis of these three countries.

mismanagement.³ As criticism of the program grew stronger within the party, Mao stepped back from the day-to-day running of the CCP. However, in 1966, he decided to reassert control by launching the Cultural Revolution.⁴ This campaign was also an economic and social tragedy for the country, leading to ten years of internal strife and virtual civil war. Millions of urban youth were sent to the countryside for "re-education" programs to purge them of capitalist or traditionalist ways of thinking, resulting in a lost generation of college students.

Following Mao's death in 1976, Deng Xiaoping became the new leader of China. After decades of disengagement with the West, Deng ushered in a new era of diplomatic relations between China and the USA, even as both sides continued to view each other with some suspicion. Deng's tenure brought with it a liberalization of the Chinese economy, opening it up to foreign investment and trade.⁵ China quickly became the fastest growing economy in the world and, over time, its free-market reforms helped lift an estimated 800 million Chinese out of poverty. China is now the world's second largest economy.⁶

Higher Education in China

Chinese families have prioritized education as a pathway to upward socioeconomic mobility for thousands of years. During the Han dynasty (which lasted from 206 BCE to 220 CE), a national examination system was set up to determine who could work in the prestigious imperial civil service. The examination required a significant amount of rote learning of Confucian texts.⁷ This practice continued until the mid-nineteenth century when the Chinese imperial army lost to the British in the First Opium War in 1842. This defeat led to growing calls within the country for an overhaul of China's education system and greater emphasis to be placed in schools on the teaching of science and technology. A steady stream of elite Chinese students began traveling to the USA and Europe (and also Japan) for higher education, with the express purpose of bringing back the new knowledge they learned

---

³ Manning and Wemheuer 2011; Thaxton, Jr. 2008.
⁴ Dikötter 2016.
⁵ Vogel 2013.
⁶ "The World Bank in China," The World Bank, accessed on January 30, 2021, www.worldbank.org/en/country/china/overview.
⁷ Peters and Besley 2018.

overseas. I mention this to highlight how Chinese students have been part of a transnational migration system for higher education for well over 100 years.

However, after becoming the PRC in 1949, higher education in China went into rapid decline. It was only after Deng Xiaoping took over the reins that the Chinese government began investing heavily in improving the quality of tertiary education within the country. Deng reestablished China's National College Entrance Examinations (known as the *gaokao*), which had been halted for ten years during the Cultural Revolution. Project 211 was launched in 1995 to improve the quality of the education and the scientific research conducted at approximately one hundred universities throughout the country.[8] Total investment in Project 211 reached USD 2.3 billion by the mid-2000s.[9] Three years later, Project 985 was announced by then-Chinese president Jiang Zemin at the 100th anniversary of the founding of Peking University.[10] The goal of Project 985 was to promote the development and improve the reputation of China's higher education system by transforming its premier institutions into "world-class" universities by the start of the twenty-first century. The project was focused on allocating resources to an initial list of nine key universities (the C9 group) to help them build new research centers, organize international conferences, and recruit international faculty and returning Chinese scholars. Even more recently, the Double First Class University Plan was launched to develop a group of forty-two universities into world-class institutions by 2050.[11]

The massive amounts of funding the Chinese government has allocated to its universities through these programs have had a significant impact. As of 2021, there are six Chinese universities ranked in the top 100 of the *Times Higher Education's* (*THE*) world university league tables. These are Tsinghua University, Peking University, Fudan University, the University of Science and

---

[8] The number "21" stemmed from the goal of preparing these universities for the 21st century while "1" represented the approximately 100 universities that were initially part of the program.
[9] Langer and Zhou 2007.
[10] Project 985 was announced in May 1998, hence the name "985" representing the year – '98 – and the month – 5 – when the announcement was made.
[11] Peters and Besley 2018.

Technology of China, Zhejiang University and Shanghai Jiao Tong University in declining order.[12]

Deng was also instrumental in restarting Chinese student migration to the West, signing agreements with the US government in the late 1970s that allowed handpicked Chinese students to pursue graduate training in the sciences and engineering in top American universities. These students were state sponsored and had strict instructions to return to China at the end of their training.[13] In the early 1980s, self-financed international student migration also began to be permitted by the Chinese state, but still with the expectation that these students would return to China at the end of their studies.

During the 1990s, the westward migration of Chinese students for graduate training continued to increase with the further opening of the country.[14] (Most of my Chinese interviewees left China in the 1990s and after.) However, the vast majority of Chinese students opted to stay in the USA at the end of their studies, rather than return home.[15] A US National Science Foundation (NSF) study reported that, 96.5 percent of Chinese PhD students who received their PhDs between 1995 and 1999 indicated they had definite plans to stay in the USA after completing their doctoral training (as compared to 58.8 percent of Taiwanese students).[16]

Starting in the late 1990s, and into the early twenty-first century, a small but increasing number of US-educated Chinese graduates started returning home.[17] Some of these return decisions were fueled by initiatives kickstarted by local and central Chinese governments to lure overseas-educated Chinese students back home after graduation. One of these initiatives was the Hundred Talents Program which was started in 1994 to encourage Western-trained Chinese academics and scientists, known as *haigui* (or "sea turtles"), to return and turn China's long-standing brain drain into a "brain boomerang."[18] Doctorate holders with at least four years of overseas postdoctoral research

---

[12] "Best Universities in China," *Times Higher Education* World University Rankings, September 4, 2020, accessed on December 18, 2020, www.timeshighereducation.com/student/best-universities/best-universities-china.
[13] Zweig and Chen 1995.
[14] Saxenian 2002.
[15] Xiang and Shen 2009. The Tian'anmen Square Massacre in 1989 led to many overseas Chinese students choosing to settle down in the West.
[16] NSF 2006: figure 6-9.
[17] Saxenian 2002; Li et al. 2018.
[18] Nature Publishing Group 2015; Wong 2008.

experience were eligible to apply for the program which led to a faculty position at one of the Chinese Academy of Sciences (CAS) institutes. In this manner, the scientist migration system between China and various advanced economies in the West began to change, with growing numbers of Chinese scientists opting to return.

The rapid development of China's economy during this time period fueled the emigration of ever more Chinese students whose families could now afford to privately fund an overseas education for their children. But this booming economy also encouraged these students' return to China at the end of their studies to take advantage of the growing job opportunities in the country. In addition, foreign qualifications were seen by Chinese families as a way to distinguish their children from the hordes of other applicants for top jobs in the country. In 2013, 413,900 Chinese students went abroad and 353,500 returned.[19] To further increase the incentives for Chinese-born, Western-trained doctorate holders to return home, China launched its Thousand Talents Program in 2008. While the program had different tracks to attract both native and nonnative highly skilled individuals to China, one track was dedicated to recruiting "junior talent" who were under forty years of age. Both the Hundred Talents and the Thousand Talents Programs led to the cumulative return of thousands of Chinese scientists over the years. Most of the Chinese returnees I interviewed had come back under the auspices of these programs, while several nonreturnees and halfway returnees had taken up part-time appointments in Chinese universities that required them to spend only three or so months of the year in China, while living in another country the rest of the year.

Despite the recent increase in postdoctoral training capacity at home, the Chinese government continues to directly and indirectly encourage its best minds to pursue postdoctoral training overseas. As mentioned earlier, in order to qualify for Scheme A for overseas "young talent" under the Hundred Talents Program established by CAS, applicants must possess at least three to four years of continuous overseas scientific research experience, which can also include overseas postdoctoral training.[20] (Critically, however, the doctoral degree that these overseas talents must possess does not need to be from an overseas

---

[19] Li et al. 2018; Kermani and Zhou 2007.
[20] "Global Recruitment of Pioneer 'Hundred Talents Program' of CAS," Chinese Academy of Sciences, December 4, 2015, accessed on June 24, 2021, http://en

institution.) Several of my younger Chinese interviewees reported that they timed their return to coincide with their completion of four years of postdoctoral training in the USA. The perceived privileging of Western (and particularly, American) postdoctoral qualifications when filling academic positions in China, coupled with the growing shortage of well-funded academic positions in China, encourages Chinese scientists-in-training to seek postdoctoral appointments overseas as a means of distinguishing themselves from the locally trained competition.[21] The remuneration levels of postdoctoral trainees in China are also reported to be low, pushing aspiring scientists to head to the West where they might not earn much in absolute terms, but still relatively more than they would have earned back home.[22] Taken together, these factors can encourage some Chinese aspiring scientists to complete their PhDs in China and then go to the West for four years of postdoctoral training before returning home to take up a full professorship.

Science and Technology in China

China has a long history of scientific innovation.[23] The compass, abacus, shadow clock, gunpowder and paper are just some of the inventions credited to the ancient Chinese. Many of these inventions diffused to the West, due to China's trade with other parts of the world, highlighting how the circulation of knowledge has not always been in a West-to-East direction.[24] In the sixteenth and seventeenth centuries,

---

glish.cas.cn/newsroom/archive/jobs_archive/job2015/201512/t20151204_157107.shtml.

[21] Zeng 2008; Xiang and Shen 2009.

[22] Zeng 2008; Gallagher 2013; Zeithammer and Kellogg 2013. This does not necessarily conflict with Stephan et al.'s (2013) finding that postdoctoral salary levels were not a significant determinant of posdoctoral trainees' destination choice. They write that survey respondents still often chose to go to the USA even though postdoctoral salary levels in the USA are not the highest in the world. However, at the regional level, postdoctoral salaries in Western countries are higher than what is available in most Asian countries, particularly India and China.

[23] The most comprehensive history of Chinese science and technology is available through Joseph Needham's magisterial volumes *Science and Civilization in China* published by Cambridge University Press.

[24] Needham 1954–2005.

a reverse flow of scientific knowledge occurred as Jesuit missionaries introduced Western innovations in mathematics and astronomy to late imperial China.[25] In the second half of the nineteenth century, Protestant missionaries also carried Western scientific knowledge to Qing China. Many classic Western textbooks in mathematics, medicine, botany and other sciences were translated into Chinese during this time period.[26]

After the defeat of China by Japan in the First Sino-Japanese War in 1895, Japan rose in status as the harbinger of modernity and science in the minds of the vanquished Chinese. Modern science was understood through the lens of Japanese interpretations of Western science. During this time, growing numbers of Chinese students went to Europe, the USA and also Japan, for further studies in modern science. Historian Benjamin Elman estimates that more than 10,000 Chinese studied in Japan from 1902 to 1907 and that the majority returned to China to take up influential positions in the civil service, further cementing Japanese influence on the study of science in China.[27] I bring up this point to highlight how scientific remittances have been an integral part of the brain circulations of scientists-in-training for well over a century, and that the directionality of these remittances have not always been limited to West-to-East flows.

During the Great Leap Forward and the Cultural Revolution, scientific research in China fell into decline. When Deng Xiaoping took over, he launched the 863 Program to stimulate internal investments in science and technology so that the country would not need to rely so heavily on foreign technologies in the long run.[28] The program identified seven priority areas for investment, including biotechnology. As part of this initiative, central funds were allocated to build university laboratories for education and basic science research. These efforts led to a large increase in the number of Chinese universities offering life science degrees and the number of Chinese students pursuing these degrees. By 2004, there were more than 150,000 undergraduate students enrolled in biology-related programs in China, representing a threefold increase in students since 1997.[29]

[25] Elman 2006.
[26] Elman 2006.
[27] Elman 2006:198.
[28] Xi and Zhang 2010.
[29] Langer and Zhou 2007.

Bioscience Research in China

By the time the country's seventh Five-Year Plan was introduced for the period 1985–1990, biotechnology was identified as the country's top priority area and funding was allocated to establish national research laboratories and centers focused on the life sciences. One of the reasons why biotechnology was selected was because, with a population of one billion people, China needed the innovations that a thriving biotechnology sector could produce, including high-yield, insect-resistant crop variants, as well as vaccinations and antibiotics for a range of diseases.[30] In 1988, the National Torch Program was announced with the goal of developing and commercializing technology (and particularly, biotechnology) through the establishment of Technology Business Incubators, "Pioneer Parks" for returning Chinese scientists, and University Science Parks.[31] These parks provided space, equipment, finance and training to aid in the industrialization of scientific inventions and discoveries, helping scientists convert their innovations into commercially profitable businesses.[32]

Biotechnology has remained a key investment area in all subsequent Five-Year Plans, with R&D spending allocated to this sector increasing with each plan.[33] Biopharmacy, bioagriculture, bioenergy and biomanufacturing are all subfields identified by the Chinese state as priority areas. In the twelfth Five-Year Plan (for 2011–2015), the government allocated USD 1.7 trillion to these efforts, with USD 1.5 billion focused on new drug development.[34] Twenty biotech zones were also set up across the nation to boost innovation and industrialization in this sector. The thirteenth Five-Year Plan (2016–2020) set a target of RMB 271 billion in spending for R&D.[35] As part of this plan, new funding lines for stem cell and translational research were also established.

These ambitious efforts have largely paid off. In 1997, there were 200 biotechnology companies in operation in China; by 2000, this number was 600, and it increased again to 900 in 2005. In that year,

---

[30] Van Brunt 1988.
[31] Xi and Zhang 2010.
[32] "Torch Program in the Past 15 Years," China.org.cn, September 17, 2003, accessed on March 25, 2020, www.china.org.cn/english/2003/Sep/75302.htm.
[33] Xia 2014.
[34] Hambley 2010.
[35] Cyranoski 2016.

the total sales revenue of these companies amounted to USD 2.4 billion. Simultaneously, there was growing recognition that China's biotech industry has begun shifting from solely producing generic biological products, to also developing more sophisticated biotech drugs, demonstrating that Chinese firms were beginning to move up the biotech value chain.[36]

Present and Future Challenges

Four factors have been identified as contributing to China's rapid ascent in global scientific circles.[37] These are China's large population and human capital base, a labor market that favors academic meritocracy, a large diaspora of Chinese-origin scientists some of whom can be enticed to return home, and its centralized government willing to invest significant sums in science research. Specifically related to bioscience, Hepeng Jia (2011), writing in *Nature Biotechnology*, lists two additional factors that facilitated China's growing standing in the global biotechnology sector. The first was the economic downturn in the USA and Europe, following the 2008 financial crisis, which led to a decline in funding rates in the West for basic scientific research. The second was the booming development of the Chinese pharmaceutical sector. This latter factor has been partly driven by the creation of R&D outposts in China by large multinational corporations, like GlaxoSmithKline (GSK) and Merck, for offshore drug development.[38] The allure of China's large domestic market for healthcare and other biotech products is a key factor attracting these multinational pharmaceutical companies to invest in China.[39]

Still, the Chinese government is the primary driving force behind the country's rise in the global biotechnology sector.[40] This state-funded growth has been critiqued for inadvertently fostering the underdevelopment of the country's venture capital industry, resulting in a weaker link between the scientific research sector and industry, as

---

[36] Kermani and Zhou 2007.
[37] Xie et al. 2014.
[38] Hambley 2010; Goodall et al. 2006.
[39] Frew et al. 2008. In this regard, China is a direct competitor of Singapore which also positions itself as an ideal base for multinational corporations' R&D centers looking to serve the lucrative and growing Asian market.
[40] Xia 2014.

well as an insufficient focus by academic scientists on intellectual property development, protection and commercialization.[41] This is an issue that other countries among my four case studies are dealing with as well.

Given the tight control that the Chinese state has over the funding of research in the country, there are also ongoing questions about the research grant allocation process. In 2014, the *Economist* reported that significant rates of fraud were uncovered in research grants managed by the Chinese Ministry of Science and Technology. Critics argued that membership in CAS and other grant-distributing bodies was too often linked to whom one knew, rather than the quality of one's research. Partly in response to this criticism, the academy changed its membership criteria in 2016, requiring all future members to be nominated by other academics or their institutions, and for their membership to be voted on by all current members. This is in contrast to the earlier approach, where nominations could be accepted from government ministries, the CCP and the army, and where voting was restricted to a small group of handpicked CAS members.[42]

As a sign of the struggle between established and newly rising members of the global scientific field, scientists outside of China have also raised questions about the validity and quality of published research coming out of the country. These concerns are partly linked to Chinese universities' incentive structure, which is geared toward rewarding the quantity of a scientist's publications and their citation rates. Publication in high impact-factor science journals also garners Chinese scientists large cash bonuses.[43] Outside observers have highlighted how a single-minded pursuit of publications in high impact-factor journals can create perverse incentives to doctor data to make a study's results appear more significant than they actually are. In 2015, the respected medical journal *The Lancet* published an editorial calling on the Chinese government and leading research institutions in China

[41] Zhang et al. 2010; Kermani and Zhou 2007.
[42] *Economist* 2016.
[43] Davis 2011. *World Education News and Reviews* reports that, in 2016, a science article published in a Western journal would receive an average cash reward of more than USD 40,000, while the author of an article in the very top science journals (such as *Nature*) could receive as much USD 160,000 (Mini Gu, Rachel Michael and Claire Zheng, "Education in China," World Education News and Reviews, December 17, 2019, accessed on March 9, 2021, wenr.wes.org/2019/12/education-in-china-3).

to change how Chinese researchers are evaluated and rewarded and, separately, how they are punished for academic misconduct, noting that the current system encouraged China's scientists to cut corners in their research process.[44] An increase in the number of retracted Chinese-authored articles containing false claims continues to muddy China's reputation within global scientific circles.[45] In my interviews, however, scientists in China and my other Asian case countries also claimed that there was a systemic bias against Asia-based scientists when it came to the peer review of their article submissions, hurting their chances of publication.

China's response to the Covid-19 pandemic is an example of both its strengths and ongoing weaknesses in bioscience. When the disease first emerged in Wuhan in late 2019, the official bureaucratic response was to deny the existence of a public health problem and scapegoat local doctors and epidemiologists who raised concerns. However, once the Chinese state accepted the severity of the situation, the government response was swift, all-encompassing, and largely successful.[46] As early as January 2020, Chinese researchers identified the genetic sequences of the Covid-19 pathogen. Hundreds of clinical trials were approved to test various treatment procedures. Five different vaccine development routes were attempted and, by the end of 2020, three Chinese companies had filed for domestic regulatory approval for a mass rollout of their vaccines. As an act of scientific diplomacy, China has framed the Sinovac vaccine, one of its domestically produced vaccines, as a global public good aimed at helping developing countries. However, questions continue to be raised about China's lack of data transparency regarding the results from its clinical trials and even the origins of Covid-19 in Wuhan.

---

[44] *Lancet* 2015.
[45] Davis 2011; *Lancet* 2015; *Economist* 2013, 2016. It should be noted that China is not unique in its incentive structures prioritizing publications in high impact-factor journals or in its experiences with academic malfeasance. Data fabrication scandals have occurred in Japan (in 2014) and South Korea (in 2004 and 2005) as well, speaking to the institutional and personal pressures on scientists in many countries to produce stellar results. It also happens that these cases of academic misconduct in Japan and South Korea involved life scientists as well.
[46] Zhang et al. 2020.

India

India is the second-most populous country in the world. With the Himalayas to its north, and the Indian Ocean, the Arabian Sea and the Bay of Bengal on its southern, western and eastern flanks respectively, a diverse set of cultures and languages developed within the Indian subcontinent over millennia. For much of this period, sea and overland trading connections with other parts of Asia and Africa existed. Starting with the arrival of Portuguese seafarer Vasco da Gama in southern India in 1498, the subcontinent also became increasingly exposed to European influence. Beginning in 1757, large areas came under British East India Company rule, followed by direct British rule from 1858. In 1947, India gained independence from the British and became a constitutional republic. While it followed a development strategy of import substitution and state-driven socialism in the initial decades after independence, the country loosened its economic controls in the 1990s, leading to rapid foreign direct investment (FDI) and a burgeoning middle class.[47] Though India is still largely an agrarian society and a developing country, its longstanding investments in engineering education have resulted in the country developing the largest IT offshoring industry in the world.[48] At the same time, India's pharmaceutical industry has become the world leader at producing generic drugs. Founded in 1966, the Serum Institute of India is the world's largest vaccine manufacturer and is also producing the Oxford/AstraZeneca Covid-19 vaccine. The Indian government hopes to leverage its competitive advantage in these sectors to enhance its status in the global scientific field more broadly.

Higher Education in India

After China and the USA, India's higher education system is the third largest in the world in terms of student numbers.[49] The University Grants Commission is the domestic organization responsible for the promotion, coordination and funding of higher education in India.[50] It

---

[47] Chittoor et al. 2008.
[48] Xiang 2007.
[49] Altbach 2005. As of 2019, there were 37.3 million higher education students in India, according to the All India Survey on Higher Education.
[50] See University Grants Commission, www.ugc.ac.in/.

oversees the large number of higher education institutes that operate in the country – 1,043 universities, 42,343 colleges, and 11,779 standalone institutions as of December 2020.[51] Roughly 70 percent are in the private sector.[52] International higher education expert Philip Altbach (2014) describes the higher education system in India as having "islands of excellence in a sea of mediocrity." His remark was aimed at the generally low quality of higher education in India, where only a select few institutions (such as the storied Indian Institutes of Technology, or IITs as they are informally called) are internationally recognized. The IITs were established by the central government in the 1950s and 1960s. They sit outside the structure of traditional universities, and enjoy little or no involvement by local governments.

In contrast, most of India's public higher education sector faces the following challenges: insufficient government funding;[53] the lack of a consistent language of instruction across the country, which undermines intra-India migration to the most competitive educational institutions where English is the medium of instruction;[54] uneven planning and coordination across higher education institutions, which are primarily governed at the state level;[55] uneven faculty quality; and high levels of bureaucracy and inertia.[56] Regarding science education in India, further systemic flaws include a rote-based approach to teaching science, which discourages a culture of innovation,[57] and low wages and low

---

[51] "AISHE 2019–20," All India Survey on Higher Education, accessed on June 25, 2021, www.education.gov.in/sites/upload_files/mhrd/files/statistics-new/aishe_eng.pdf.
[52] Sarukkai 2015.
[53] Agarwal 2006.
[54] Altbach 2014. This is a particularly acute problem when it comes to graduate education in the sciences as students from rural and semi-urban areas are more likely to have attended schools where the medium of instruction was not English and so may find it difficult to pursue further study in STEM fields, where English proficiency is necessary (Sarukkai 2015).
[55] Altbach 2014.
[56] Altbach 2014.
[57] Chaurasia 2016. Many scholars have noted the shortage of qualified science teachers in primary and secondary schools in India. Taking a ground-up approach, Rajesh Kochhar, Professor at the Indian Institute of Science Education and Research (IISER), Mohali, and former director of the National Institute of Science Technology and Development Studies (NISTADS), New Delhi, argued that the absence of world-class scientists in India is not as worrying as the absence of science teachers for Indian children (Jayaraman and Priyadarshini 2012).

status associated with a career in science.⁵⁸ No Indian university ranks in the *THE* World University Rankings top 100. The Indian Institute of Science (IISc), which was established in Bangalore in 1909, is considered the top university in the country but only ranks in *THE*'s top 300 universities of the world.⁵⁹ To encourage more young people to pursue an education in the sciences in India, seven Indian Institutes of Science Education and Research (IISER) were established around the country in the 2000s, modelled after the world-famous IITs. These new universities allow students to enroll in joint undergraduate and master's programs in various scientific fields, and potentially continue to a PhD program. Despite their infancy, these universities have developed an impressive track record after educating hundreds of master's and doctoral students in India, and rising in the *Nature Index* rankings.⁶⁰

When it comes to postdoctoral training, India has also only recently begun upgrading its scientific training infrastructure.⁶¹ For a long time, few postdoctoral training positions were available in the country and those that did exist offered low monthly stipends. In 2009, there were just over 300 postdoctoral scholars across all major bioscience departments and research institutes in India.⁶² Likewise, only 11 percent of researchers at major and minor universities in India had completed their postdoctoral training in India, while almost 60 percent had received their training in a single overseas country: the USA.⁶³ The central government has since established several new postdoctoral fellowship schemes with higher salary rates.⁶⁴ However, these fellowships do not yet match the prestige associated with a Western qualification and, as a result, these

---

[58] Indian high school students who are considering a career in science may be daunted by the fear that pursuing scientific research will not provide them with sufficient salary to support a family – due to the low financial returns and long timespan required for the completion of formal studies (Ghosh and Kshitij 2016 citing Agarwal 2009; Sarukkai 2015).

[59] "Indian Institute of Science," *Times Higher Education* World University Rankings, accessed on January 31, 2021, www.timeshighereducation.com/world-university-rankings/indian-institute-science.

[60] *Nature* 2016.

[61] Scientific Advisory Council to the Prime Minister 2011.

[62] Vale and Dell 2009.

[63] National Institute of Science, Technology and Development Studies (NISTADS) 2009:33.

[64] Scientific Advisory Council to the Prime Minister 2011.

postdoctoral fellowship slots have reportedly not been fully subscribed.[65]

Like China, India has its own long-standing international student migration system involving the West. Several of the leaders of India's independence movement had trained in the UK and then returned to India, arguing that British ideals of democracy should be extended to India. Even after independence, this westward student migration (particularly to the UK) remained the preserve of upper-class Indian families.[66] But since the 1990s, the growing upper-middle-class and shrinking family size has allowed more Indian families to consider sending their children overseas. The OECD estimates that 223,000 Indians were studying abroad in 2011, with 90 percent of them in OECD countries.[67] The USA has now surpassed the UK as the top destination for Indian student migrants. In 2012, there were close to 100,000 Indian students studying in the USA, as compared to only 22,000 in the UK. Alternative destinations in the West, including Australia and Canada, and destinations in Asia itself, particularly Singapore, are also growing in popularity.[68]

As with Chinese students, many Indian students see an overseas degree as a chance to distinguish themselves from the crowded field within India, as well as a way to settle permanently overseas.[69] India's Economic Survey of 2018 found that, of the more than 100,000 Indian-born doctorate holders living and working overseas, over 91,000 lived in the USA alone. Unlike overseas-educated Chinese students, the number of overseas-educated Indians returning home remains small, though it is increasing.

Between 2007 and 2012, the number of returning Indian scientists was only 243. This number climbed to 649 between 2012 and 2017.[70]

[65] Vale 2014.
[66] Khadria 2015.
[67] "How is International Student Mobility Shaping Up?" *Education Indicators*, July 2013, accessed on December 17, 2020, www.oecd.org/education/skills-beyond-school/EDIF%202013–N%C2%B014%20(eng)-Final.pdf.
[68] King and Sondhi 2018; Ortiga et al. 2018; Chanda 2015. As I discuss in the Singapore section of this chapter, part of Singapore's human-capital strategy involves wooing young talent from neighboring countries like China and India into its educational institutions through scholarships and fellowships. I was a direct beneficiary of this strategy when I won a pretertiary scholarship for Indian students to study in Singapore.
[69] King and Sondhi 2018.
[70] Bhattacharya 2018.

Fellowship programs with generous research packages to entice Indian scientists to return are partly credited for this doubling of return statistics. These fellowships include the Ramanujan Fellowship Scheme, the Innovation in Science Pursuit for Inspired Research (INSPIRE) faculty scheme from India's Department of Science and Technology, and the Ramalingaswami Re-entry Fellowship from the Department of Biotechnology (DBT). The latter fellowship offers research grants for a period of five years to returning scholars. The Wellcome Trust-DBT Indian Alliance also gives more than USD 200,000 to outstanding early-career returning scientists to help them set up their laboratories in India, and even more to senior scientists.[71]

Science and Technology in India

Indians have long prided themselves on their country's many ancient contributions to the body of universal scientific knowledge, particularly in the realms of astronomy, medicine and mathematics.[72] However, the longstanding orientalist view was that Indian science had withered during the seventeenth and eighteenth centuries, until the British arrived and began introducing Western ideas of scientific practice.[73] George Basalla's (1967) ideas about the diffusion of Western science to the rest of the world is an example of this historical worldview that imagines a unidirectional flow of Western scientific ideas to the *tabula rasa* of the Western colonies in Asia, Africa and elsewhere. The propagation of Western and colonial science was viewed by British Raj scientists and administrators as part of their "civilizing" mission, even as it was also directly tied to profit motives. But the conflation of Western science and modernity was also absorbed by many Western-educated Indian elites who saw themselves as agents of progress (and eventual independence) through their promotion of modern scientific practices and knowledge.[74] Independent India's first prime minister, Jawaharlal Nehru, was one such avid supporter of

---

[71] "Basic Biomedical Research Fellowships," IndiaAlliance-DBT welcome, accessed on June 25, 2021, www.indiaalliance.org/fellowshiptype/basic-biomedical-research-fellowships.
[72] Baber 1996; Arnold 2000; Dhawan, Gokhale and Verma 2005; Subbarayappa 2013.
[73] Arnold 2000; Subbarayappa 2013.
[74] Prakash 1999.

science, seeing it as a developmentalist tool for postcolonial nation-building but also as an ideological tool for staking a nationalist claim regarding ancient India's contributions to scientific advancement.[75] For Nehru and nationalist leaders in other former colonies around the world, modern science was "state science—science conducted for the people but at the direction and discretion of the state" (Arnold 2013:366).

In 1958, the Indian government passed its Scientific Policy Resolution, a high-level policy statement that outlined the need to harness science and technology as a tool for national transformation.[76] This statement demonstrates a very different perspective on scientific values compared to Robert Merton's scientific ideals of universalism and disinterestedness (described in Chapter 1). For many developing countries like India, the role that scientists were expected to play in national development presupposed a starkly different relationship between scientists, the knowledge they produced, and the government funding agencies that supported their research. Rather than pursuing science for science's sake, there was an understanding that investments in science were expected to directly benefit the nation through improvements in crop yields, reductions in mortality rates, increased economic output, and greater recognition and respect on the world stage – to name a few possible returns on this investment. Western countries often invested in scientific research for similar reasons: glory, profit and power. But because of their long-standing position of superiority within the global scientific field, these countries had the luxury to construct a narrative of disinterestedness around the goals of scientific research that masked the advantages that automatically accrued to them as a result of the concentration of scientists and research resources within their borders.[77] However, as Western countries' position in the global scientific field comes under threat because of large investments in science by Asian

---

[75] Arnold 2013.
[76] Accessed on December 17, 2020, https://indiabioscience.org/media/articles/SPR-1958.pdf.
[77] George Basalla (1967) represents such a view of Western countries' scientific endeavors as largely motivated by a neutral desire to gain more systematic knowledge about the natural world, eliding more mercenary or political motivations. More recent STS scholars acknowledge that Western countries are fully aware of the economic and geopolitical benefits that come with maintaining their scientific lead in the international arena (Xie and Killewald 2012; Jasanoff 2005).

countries, the narrative of universalism in science is becoming somewhat harder to maintain.[78] As I show in subsequent chapters, this tension between established actors and aspiring entrants in the global scientific field is growing, and scientists who were trained in the West but now work in Asia may find themselves caught on one side or the other of this debate about the appropriate goals and values of their research.

Bioscience Research in India

While there were already pockets of scientific excellence in physics, chemistry and mathematics at the time of India's independence, there was only limited research being conducted in the biological sciences.[79] This began to change in the 1960s when the Indian government initiated a "green revolution" to gain food independence and help feed its people through modern agricultural techniques and high-yield varieties of grains.[80] The so-called father of the green revolution in India, M. S. Swaminathan, saw biotechnology as the pro-poor, pro-jobs and pro-nature answer to questions of development for former colonies.

From the 1970s onwards, the Indian government invested further in the life sciences. Under the auspices of the Council of Scientific and Industrial Research (CSIR), which is India's leading R&D government organization, several national research institutes in stem cell research, cell science, translational health science, agri-food biotechnology, and biomedical genomics were established. Some of the more prominent ones were the Centre for Cellular and Molecular Biology (CCMB) set up in Hyderabad in 1977 and the National Centre of Biological Sciences (NCBS), which was formally established in 1991 in Bangalore. In 1986, then-prime minister Rajiv Gandhi established the DBT within the Ministry of Science and Technology, with the mandate to develop the necessary scientific infrastructure for India to compete on the international stage in the life sciences.[81]

In the present day, India has established a niche for itself in the international pharmaceutical industry with its production of low-priced

---

[78] Marginson 2021.
[79] Dhawan, Gokhale and Verma 2005.
[80] Visvanathan and Parmar 2002.
[81] *Asia Pacific Biotech News* 2004.

generic drugs and vaccines for the world at large.[82] The seeds for this competitive edge were first sown in 1970 when India's Patents Act introduced a new patent-law regime that restricted patent eligibility to the manufacturing processes for substances to be used as food, drugs or medicines, but not the substances themselves. This provision was introduced into the Patents Act to ensure that India would not become dependent upon expensive imports of patented Western medicines.[83] As a result, an innovative pharmaceutical industry developed within India, becoming expert in reverse engineering the manufacture of essential medicines at a low cost. India's ascension to the World Trade Organization in 1994 led to increased pressure on the country to change its patent laws to become more aligned with those in the West. In 2005, India finally amended its laws to allow for product patents, and not just process patents, in the pharmaceutical sector. However, it also included a provision that new versions of existing products could not automatically qualify for an extension of their original patent if the new version of the product failed a novelty test. This provision has allowed India's pharmaceutical sector to continue to thrive as a source of low-cost generic drugs even as the sector has engaged in significant internationalization.[84] Indian research institutes and pharmaceutical companies also partner with foreign companies to run large-scale clinical trials for vaccines and produce them in-country (as in the case of the Oxford/AstraZeneca Covid-19 vaccine). They also conduct research to develop their own indigenous vaccines, such as Covaxin produced by Indian company Bharat Biotech. The medical biotechnology and health sectors – which include human genetics, genomics and vaccine research – are seen as having great potential for monetization because of these early investments.[85]

However, outside of these sectors, India does not yet have a large global footprint in basic bioscience research.[86] Several initiatives have been launched to address this gap. IndiaBioscience (IBS), an organization promoting life science research in India, was set up to strengthen

---

[82] *Nature* 2015a; Chakraborty and Agoramoorthy 2010; Chakma et al. 2010; Chittoor et al. 2008.
[83] Chittoor et al. 2008.
[84] See Chakma et al. (2010) for a case study of one such homegrown Indian vaccine developer. See also Chittoor et al. 2008; Frew et al. 2007.
[85] Chaturvedi 2005; Goodall et al. 2006.
[86] Vale and Dell 2009; *Nature Materials* 2009.

the network of life science researchers and students in the country. Since 2009, IBS has helped organize the DBT-funded Young Investigators' Meeting, which is held annually to bring together promising young Indian life science researchers so they can meet with senior faculty mentors, representatives from grant-funding agencies, and editors of top life science journals to share their research and receive career advice.[87] But India has a long way to go before its bioscience research sector will match what is available in countries like China or even in smaller (but richer) nations like Singapore and Taiwan.

Present and Future Challenges

As a result of the many problems with India's higher education sector, research institutions in the country often face difficulties with recruiting local scientific talent with the requisite skills. In an infographic released by *Nature* in 2015, India had a reported 200,000 full-time researchers for its population of 1.3 billion people, amounting to 4 researchers per 10,000 of the labor force, as compared to the United States' 79 researchers per 10,000 of the labor force.[88] Many undergraduate science programs in regional universities and colleges throughout the country still lack sufficient hands-on training and scientific equipment, leaving Indian students underexposed to graduate research opportunities and under-equipped to find jobs in the biotechnology sector.[89]

For those students who seek a career in scientific research, the allure of the West is strong as India's scientific research system remains underdeveloped on many fronts – except in a handful of top research institutes – with limited slots for doctoral students. The Indian government has attempted to reduce the volume of brain drain and to entice overseas-trained scientists to return, but more can be done – especially in terms of the quantum of funding made available for R&D activities. While overall R&D investment rates have increased in the last few decades, critics have pointed out that India "devotes less than 1 [percent] of its GDP to R&D, which puts it far behind emerging nations such as China and Brazil, as well as the established economies of the US

---

[87] "Young Investors' Meeting Series," IndiaBioscience, accessed on December 23, 2020, https://indiabioscience.org/yim-series.
[88] Van Noorden 2015.
[89] Basu 2007.

and Europe."⁹⁰ Research investments are reported to have hovered between 0.6 and 0.7 percent of India's GDP for the first two decades of the twenty-first century.⁹¹

As I noted in Chapter 1, rebalancing the Asian scientist migration system, which currently largely benefits Western countries, requires significant investments by Asian governments. The Indian government has tried to improve its funding track record. For Indian companies and entrepreneurs, the Technology Development Board in the Department of Science and Technology and the New Millennium Indian Technology Leadership Initiative of the CSIR are two new government sources of grant funding.⁹² In 2009, the Indian government also partnered with several biotech companies in a cost-and-risk sharing scheme as part of a pioneer Biotechnology Industry Partnership Program. The average size of grants coming from the DBT and CSIR is increasing, such that researchers can spend less time applying for multiple grants and focus more on their research. Still more can be done upstream to improve the research environment in India.

Singapore

Singapore is a Southeast Asian island nation situated just below the Malaysian peninsula, with Indonesia to its south. As of 2021, it has a population of just under six million, with the majority ethnic group being Chinese, followed by Malays, Indians, Eurasians and various other ethnic Asian groups, in that order.

For centuries, Singapore was a common port of call on Southeast Asian, Chinese and Indian shipping routes. In 1819, Sir Stamford Raffles of the British East India Company selected Singapore to be a free port after gaining control of the island from the local Malay ruler. Singapore was ruled by the British for over a hundred years and, during that time, its ethnic make-up changed drastically as tens of thousands of Chinese and Indian immigrants moved to the island seeking work. After World War II ended, there was a growing push in Singapore for independence. In the 1950s, Singapore was made self-governing and then, in 1963, it joined the Federation of Malaysia. But

---

⁹⁰ *Nature* 2015a.
⁹¹ Business Today 2019.
⁹² *Asia Pacific Biotech News* 2004.

this arrangement was short-lived as Singapore was forced to secede from Malaysia and become an independent nation in 1965.[93]

Singapore's unexpected independence engendered a deep-seated anxiety about the country's long-term viability, given that it lacked any natural resources and had a largely uneducated and multiethnic immigrant population with little sense of a shared national identity. Under pressure to unify and stabilize the country, the ruling People's Action Party (which continues to dominate Singapore politics) adopted aggressive developmental strategies based on the internationalization of the economy, which resulted in rapid economic growth. In the present day, Singapore has one of the highest GDP per capita ratios in the world, a well-established reputation as a global city, but also a rapidly aging population. While the economy is still reliant on its airport and seaport, Singapore has also invested heavily in its knowledge economy in recent years to diversify its economic base.[94] It is also highly dependent on low- and high-skilled migrant workers to fill severe shortages at both ends of the labor spectrum – much more so than my other three country case studies. This injects significant diversity into its population but also leads to tensions between foreigners and locals.[95] Later in the book, I discuss the challenges Singapore faces as it tries to balance a desire for diversity in its workplace with the principle of meritocracy.

Higher Education in Singapore

Prior to independence, students in Singapore who wanted to pursue higher education in the sciences had limited local options. The University of Malaya had two branches – one in present-day Singapore and the other in present-day Malaysia, and English was the language of instruction at the university. Ethnic Chinese students in Singapore who had been educated in a Chinese dialect could enroll in the privately funded Nanyang University, which was founded in the early 1950s with Chinese as its language of instruction.[96] In the early 1960s, the Singapore branch of the University of Malaya was renamed the University of Singapore. Nanyang University and the University of

---

[93] Hack and Margolin 2010; Kwa et al. 2019.
[94] Thampuran et al. 2017.
[95] Ng 2013.
[96] Wong 2000.

Singapore were merged in 1980 to become the National University of Singapore (NUS) and all university instruction in the country was henceforth carried out in English.

Independent Singapore's second university – Nanyang Technological University (NTU) – was founded in 1991. There are currently six local universities in Singapore, with NUS and NTU being the most established. Both are in the *THE* World University Rankings top 100 for 2021, with NUS at 25th place and NTU at 47th.[97] NTU is also ranked first in the world in the *THE*'s Young University Rankings for 2021.[98] Both NUS and NTU are comprehensive research-intensive universities with their own medical schools and respected faculties of engineering and science, among other disciplines. They attract a significant number of international students, drawn largely from other parts of Asia, and they also have a high proportion of foreign faculty on staff. Both also have partnerships with multiple foreign universities to further expand their international standing, offer joint degrees, and increase the volume of transnational research collaborations.[99]

Singapore relies on a long-standing government scholarship system to send its top students to universities in the West for higher education and training. These students are then required to return to Singapore to fulfill a multiyear service bond in local universities or research institutes as researchers or teaching faculty, or in various wings of the country's civil service. Singapore's Ministry of Education also sponsors high-achieving students from China, India and various ASEAN countries to study in Singapore for high school and/or university as a brain gain strategy, with the hope that some of these students may choose to naturalize and settle down in Singapore.[100] The

---

[97] "Best Universities in Singapore," *Times Higher Education*, September 16, 2020, accessed on January 31, 2021, www.timeshighereducation.com/cn/student/best-universities/best-universities-singapore.

[98] "Nanyang Technological University, Singapore," *Times Higher Education*, accessed on June 26, 2021, www.timeshighereducation.com/world-university-rankings/nanyang-technological-university-singapore.

[99] Yale-NUS College, for instance, where I teach, is a four-year liberal arts college within NUS that was set up in partnership with Yale University. NUS also partnered with Duke University to set up its second medical school (built along US graduate medical school lines), while NTU partnered with Imperial College London to set up its own medical school.

[100] This was how I first came to Singapore as a sixteen-year-old student from India. I was assigned to a junior college in Singapore on a two-year scholarship

Singapore government has also established several scholarship programs (such as its National Science Scholarship)[101] to fund the overseas PhD studies of promising science students. These scholarships are available not only to Singaporean citizens and permanent residents, but also to noncitizen undergraduate students in Singapore who are willing to divest themselves of their existing nationality and take on Singaporean citizenship.[102] In exchange for generous funding to support their overseas doctoral training, these scholars are required to return to Singapore at the completion of their PhD programs and serve in a local research institute as a postdoctoral fellow for four years. A*STAR, the Singapore government agency in charge of science and technology research, has provided more than 1,200 such scholarships to students over the last ten years, ensuring a steady supply of young scientific talent for Singapore's research organizations.

Science and Technology in Singapore

As with most developmental states, Singapore's R&D activities in science and technology have been largely initiated and driven by the state.[103] In 1967, just a few years after gaining independence from Malaysia, the Singapore government established its first Science Council, which was tasked with making recommendations to the government on scientific and technological R&D matters – including the training and utilization of scientific manpower as well as the establishment of official relations with scientific organizations from other countries.[104] Since Singapore's first national science conference in 1968, the emphasis has always been on "how science and technology could work hand-in-hand with industry to boost economic development" (Tan 2017:2). This industry-aligned approach to scientific R&D

---

ostensibly sponsored by Singapore Airlines but, in actuality, paid for by the Singapore Ministry of Education.

[101] "National Science Scholarship," Agency for Science, Technology and Research (A*STAR), accessed on June 3, 2018, www.a-star.edu.sg/Scholarships/For-Graduate-Studies/National-Science-Scholarship-PhD.

[102] "Graduate Studies," Agency for Science, Technology and Research (A*STAR), accessed on June 26, 2021, www.a-star.edu.sg/Scholarships/for-graduate-studies.

[103] Koh and Wong 2005; Lai 2004; Parayll 2005.

[104] Tan 2017.

is still part of Singapore's long-term economic strategy, leading to a governmental emphasis on applied rather than basic science.

In 1980, the country's first science park was established near the NUS campus, with a focus on manufacturing science. However, as Singapore developed, it was no longer able to position itself as a low-cost manufacturing hub to MNCs. And so, the government decided to accelerate the transformation of the country into a knowledge economy. In 1985, Singapore's first life science research institute, the Institute of Molecular and Cell Biology (IMCB), was set up. South African biologist Sydney Brenner served as founding chairman of IMCB's scientific advisory board.[105] In 1989, IMCB entered into a fifteen-year partnership with the pharmaceutical giant Glaxo (which later became GlaxoSmithKline or GSK) for research on degenerative brain diseases.

In 1991, the Science Council was replaced by the National Science and Technology Board (NSTB). Its mission was expanded to promote R&D activities, including the establishment of research institutes, dispersal of research funding, recruitment of foreign research personnel, and implementation of programs to build research expertise in key areas of growth identified by the government.[106] The NSTB was established under the authority of the Ministry of Trade and Industry, signaling how scientific research was viewed by the Singapore state as an upstream activity that fed directly into industrial applications.

During these early years, however, Singapore never invested more than 0.5 percent of its GDP on R&D activities, largely relying on FDI to fund research.[107] Then, in the 1991 National Technology Plan, a total of SGD 2 billion (or approximately USD 1.2 billion)[108] for five years was set aside for R&D.[109] This amount translated to 0.85 percent of Singapore's GDP at the time, which was still significantly less than what other newly industrialized nations in Asia were investing.[110] With Singapore's third Five-Year Plan starting in 2005, however,

---

[105] Brenner went on to win the Nobel Prize in Medicine in 2002.
[106] "National Science and Technology Board is Formed," January 11, 1991, accessed on March 6, 2020, http://eresources.nlb.gov.sg/history/events/9addb3d8-62eb-420a-a10a-3727f790259c.
[107] Thampuran et al. 2017:39.
[108] Based on the historical exchange rate of SGD 1.743 = USD 1 from January 1991.
[109] Ministry of Trade and Industry 2006.
[110] Yeoh 2017:28.

government investment in R&D increased to SGD 6 billion, which represented 2.25 percent of its GDP. From 2005 to 2015, the government budget for R&D increased to 3 - 4 percent of GDP.[111] In 2016, the government set aside a record SGD 19 billion (or USD 13.2 billion) [112] in its Five-Year Plan for R&D. Within this amount, SGD 4 billion was earmarked for the health and biomedical sciences.[113]

Bioscience Research in Singapore

It was in the late 1990s that Singapore decided that biotechnology could be a growth industry for the country. Singapore's roadmap for biotechnology success involved leveraging its existing connections with multinational biotech corporations, remaking itself as a regional or even global biotech research hub and a gateway to the Asian market.[114] Beginning in 2000, the state laid out plans to make Singapore an attractive destination for biomedical researchers.[115] In June of that year, Singapore launched the Biomedical Sciences Initiative, with the goal of making the biomedical sector one of four key pillars of Singapore's economy, alongside electronics, engineering and chemicals.[116] In 2002, the NSTB was renamed the Agency for Sciences, Technology and Research (A*STAR) to signify the shift in focus towards a more research-oriented science industry.[117] At the same time, the Biomedical Research Council was formed under A*STAR to boost research in the life sciences.

The first half of the 2000s was focused on encouraging more pharmaceutical and biotechnology industry actors to establish offices and research centers in Singapore. In 2003, the biosciences hub Biopolis was launched as "Singapore's ecosystem of bioscience institutions" (Ong 2016). Situated just a short taxi ride from NUS, several science parks and some of the country's top hospitals, the Biopolis

---

[111] Lai 2007.
[112] Based on the historical exchange rate of SGD1.4243 = USD1 from January 2016.
[113] "Where the S$19b is going to," Business Times, accessed on December 15, 2020, www.businesstimes.com.sg/sites/default/files/attachment/2016/01/09/BT_20160109_UWRESEARCH9AG8KR_2058716.pdf.
[114] Wong 2011:4.
[115] Thampuran et al. 2017.
[116] Yeoh 2008.
[117] Chang 2003; A*STAR 2012.

campus was designed as a futuristic haven for basic and applied scientific research on the human genome, biostatistics, and drug discovery and development for Asian markets.[118] Early corporate entrants which set up R&D offices at Biopolis included Novartis, Eli Lilly and GSK.[119] Meanwhile, various local research institutes – the Bioinformatics Institute, the Bioprocessing Technology Institute and the Genome Institute of Singapore, among others – moved offices to the Biopolis campus too. A*STAR also worked to attract renowned scientists from the West to set up research labs in a Biopolis institute, by offering attractive research packages.[120] Some of the scientist "superstars"[121] who moved their labs to Singapore included Neal and Nancy Copeland (the scientist-couple known for their research on how tumors evolve and the development of drug resistance), Sir David Lane (the cancer researcher who discovered the gene linked to half of all human cancers), Alan Coleman (who was part of the team that cloned Dolly the sheep), and Edison Liu (who had previously directed the US National Cancer Institute).[122] With these superstar scientists setting up labs in Singapore, many younger scientists in the West decided to follow their lead and move to Singapore too,[123] creating significant media buzz about the prospects of doing cutting-edge research in Singapore.[124]

During this time, Singapore was also restructuring and expanding its national universities to reorient them towards research and not just education.[125] In 2001, NTU established its School of Biological Sciences. In 2005, NUS partnered with Duke University from the USA to establish the Duke-NUS Medical School – a graduate medical school issuing American-style graduate medical degrees and emphasizing clinical research as a core part of its curriculum and training.

---

[118] Ong 2016.
[119] Goh Yee, "Singapore Playing an Instrumental Role in the Life Science Sector," Singapore Biotech & Pharma Guide (pharmbiosingapore.com), accessed on March 6, 2020, www.pharmbiosingapore.com/articles/industry-articles/singapore-s-biomedical-sciences-initiative-reflections-for-the-future.
[120] Cheam 2005; Chang 2003; Lee 2005.
[121] Zucker and Darby 1996.
[122] Smaglik 2003; Copeland and Jenkins 2009.
[123] Most of the Western-born scientists I interviewed in Asia were based in Singapore.
[124] Krishna and Sha 2015; Chan 2006.
[125] Paul and Long 2016; Wong 2011.

During the second phase of the Biomedical Sciences Initiative, beginning in 2006, the government's focus shifted towards strengthening the country's capabilities in translational and clinical research.[126] Five new national research centers, focusing on various cutting-edge areas of life science research, were founded to spur research excellence in local universities. This project was headed by Singapore's National Research Foundation in collaboration with the Ministry of Education, NUS and NTU. These national research centers included the Cancer Science Institute of Singapore (established in 2008), the Mechanobiology Institute (established in 2009), and the Singapore Centre on Environmental Life Sciences Engineering (established in 2011). Meanwhile, NTU launched a new medical school in 2013 in partnership with Imperial College London.

In 2010, even greater emphasis was placed on industrial alignment to increase Singapore's returns on its investments more rapidly.[127] The government explicitly indicated that it would allocate more funds to research institutes and centers via competitive, rather than block, grants.[128] A greater proportion of national R&D funds was also reoriented towards direct economic outcomes through applied, rather than basic, research.[129] A SGD 1.35 billion (or USD 0.96 billion)[130] Industry Alignment Fund was established to encourage public researchers to work with industry partners on research projects that had matching funds provided by industry. There was also increased support for scientists to commercialize their research findings, through a doubling of the Research Innovation and Enterprise budget to about SGD 1.08 billion.

---

[126] Yeoh 2008.
[127] Fischer 2013.
[128] Block grants involve the disbursement of a large sum of money to an organization, leaving it to the organization to determine the criteria for distributing the funds from the block grant to its faculty/researchers. Block grants have the potential advantage of allowing researchers within the organization to enjoy easier access to research funds without having to spend as much time writing grant applications and competing with researchers from other organizations. However, this approach can also result in researchers being allocated internal funds for projects that are not very promising or that would not have received funding through a more competitive and open selection process.
[129] Ministry of Trade and Industry 2011.
[130] Using the historical exchange rate of SGD 1.40 = USD 1.00 from January 2010.

Naturally, not everyone was happy about this shift away from basic towards more applied research, and the 2010s saw an exodus of some of the foreign superstar scientists who had been recruited to Singapore from the West just a few years earlier.[131] This tension between researchers who want to focus on basic research and policy-makers who want to emphasize applied research for quick turnarounds is an ongoing issue in Singapore, due to its small size and sense of economic vulnerability.[132]

Present and Future Challenges

As discussed above, an ongoing challenge for Singapore has been how much to invest in basic research, where the payoffs are uncertain. In the late 2000s, there was growing dissatisfaction within government circles about its large bioscience research investments, which were not providing a return as quickly as some officials had hoped (or were used to, from their experience with other sectors of the economy that were less research driven).[133] As no industry-changing discoveries or drugs had emerged after the first few years of heavy investment, the government decided to shift emphasis to supporting more translational research and industry collaborations.

After a period of unhappiness and high personnel turnover as a result of these changes, the bioscience sector in Singapore appears to have stabilized somewhat as newly arriving researchers are more aware of the government's preference for translational research. However, basic research is still funded through dedicated pools of money provided by the Ministry of Education and the National Research Foundation.

A second issue Singapore faces is tied to its manpower shortage issues. Given its small population base, Singapore will always need to rely on foreign-born scientists, but this comes with challenges in talent

---

[131] Disease Models and Mechanisms 2012.
[132] Certainly this tension between what is known as Mode 1 and Mode 2 research exists in many parts of the world (Nowotny et al. 2001; Gibbons et al. 1994). Mode 1 processes of knowledge production – also known as basic research – largely focus on theoretical/fundamental questions established by the academic community themselves. Mode 2 knowledge production focuses on applied questions that bring in other actors – from the government to industry players to the general public – to give input on the research questions worth pursuing.
[133] Wong 2011.

recruitment and retention as I discuss in Chapters 5 and 7. When trying to recruit nonnative researchers to its shores, Singapore does not have the option of leveraging loyalty to the nation as a recruitment strategy. Foreign scientists may be less inclined to acquiesce to the government's goal to have state-funded scientific research support national priorities.[134] Expatriate scientists, who relocate to Singapore for the greater research funding opportunities it offers, may rankle at the state's underlying motivations to boost national economic development. These scientists may view the idea of a science that primarily serves a single nation as too narrow and counter to the universalistic goals of scientific research to which they were introduced in the West. Despite the Singapore government's heavy investment in nation-building, even Singaporean scientists may feel limited loyalty to their young country – believing that their personal research careers will be better served in the West, where they may pursue a wider range of research questions. And so both personal ambition and a lack of a sense of belonging to Singapore may cause scientists to be more impatient when they encounter research hurdles in Singapore. Given that scientists are relatively mobile migrants, a perceived deterioration in the country's research environment may encourage their departure. In fact, an ongoing concern for Singapore is whether foreign scientists will simply use Singapore as a stepping-stone in their research careers, leveraging its bountiful research funding to boost their research output but not committing longer term to the country and its future. This question of the ideal distribution between native-born versus naturalized versus noncitizen scientists in a research setting is a live issue that more and more countries have to grapple with as they seek to diversify their research community while maintaining their goal of science in the service of the state.[135]

Finally, Singapore is also struggling to find its niche within the global biotechnology industry. As China and India offer Western companies increased access to their domestic markets, there is an open question about the need for an intermediary state like Singapore as a gateway to the rest of Asia.

---

[134] Ong 2016.
[135] Ng 2013.

Taiwan

Officially known as the Republic of China, the nation of Taiwan is situated in East Asia, off the coast of the PRC. As of 2020, it had a population of close to twenty-four million people.[136] Inhabited for thousands of years by indigenous tribes, it was colonized by the Dutch, the Spanish and the Ming Dynasty for successive periods starting from the seventeenth century, and then ruled by Japan from 1895 until 1945. After the Japanese retreated from the island at the end of World War II, the KMT reasserted control over the territory. In 1949, the Chinese Civil War ended with the KMT losing mainland China to the CCP. KMT leader Chiang Kai-Shek withdrew to Taiwan and declared martial law throughout the island.[137] Taiwan remained under single party, authoritarian KMT rule for forty years until democratic reforms were instituted in the 1980s. By then, Taiwan had undergone substantial economic transformation largely based on export-driven growth.[138]

Taiwan's rapid industrialization began in the 1960s and 1970s as American businesses sought low-cost manufacturing centers outside the USA in response to competition from cheaper Japanese imports. Taiwan's semiconductor industry benefited from this American FDI, as well as from the return of US-trained Taiwanese engineers.[139] During this period, Taiwanese policymakers invested heavily in the country's technical infrastructure and native-born skills base, creating an attractive environment for investors from Silicon Valley. All of these initiatives helped cement Taiwan's lead in the global semiconductor industry.[140] Taiwan is now considered a developed capitalist economy.

Its geopolitical status is less sanguine, however, as China claims Taiwan under its One China Policy. Taiwan lost its membership in the United Nations in 1971, when its seat was given to the PRC. Today, Taiwan has only unofficial diplomatic relations with many countries, as China increasingly asserts itself on the world stage. A recent example of the constraints Taiwan faces is how Taiwan's response to the Covid-19 pandemic was largely ignored by the World Health

---

[136] "Statistical Yearbook of the Republic of China 2019," National Statistics, edited 2020, accessed on December 14, 2020, https://eng.stat.gov.tw/public/data/dgbas03/bs2/yearbook_eng/y003.pdf.
[137] Manthorpe 2005.
[138] Amsden 1979; Roy 2003; Kuznets 1988.
[139] Saxenian 2002; Iredale and Guo 2003; Chang 1992.
[140] Liu 1993; Mathews and Cho 2000.

Organization (where it is a nonmember), even though Taiwan's swift and effective public health response was internationally praised.[141] Such geopolitical concerns can make Taiwan a risky bet for multinational pharmaceutical corporations seeking a base for their East Asian headquarters or an R&D office.

Higher Education in Taiwan

There are many public and private universities in Taiwan, as Taiwanese families (and the Taiwanese state) place a great deal of emphasis on educational attainment and achievement as markers of social status and a pathway to upward socioeconomic mobility. However, as in China, Taiwan's approach to learning is often criticized for being fixated on rote learning, examinations and grades, and developing only a surface understanding of scientific concepts.[142]

National Taiwan University is the flagship higher education institution in the country. It is ranked 97th in the 2021 *THE* World University Rankings,[143] and is particularly well-known for its science and engineering programs. However, none of Taiwan's other universities rank in the top 200 of world universities. Their level of internationalization is also not high, with a relatively small proportion of international students and international faculty. In 2019, for instance, approximately 7.6 percent of National Taiwan University's student body were international students,[144] while 8.7 percent of its faculty were foreign.[145]

Armed with a local undergraduate degree, thousands of Taiwanese students pursue overseas graduate education each year, partly because international qualifications are a way to stand out in the crowded domestic field and also as a pathway to gain citizenship

---

[141] Chen and Cohen 2020.
[142] Wang and Liou 2017; Tsai and Kuo 2008.
[143] "National Taiwan University," *Times Higher Education* World University Rankings, accessed on June 26, 2021, www.timeshighereducation.com/world-university-rankings/national-taiwan-university-ntu accessed.
[144] "Number of domestic and foreign students, sorted by college, 1980–2019," accessed on February 1, 2021, http://acct2019.cc.ntu.edu.tw/acct2019e/acct1/32.pdf.
[145] "Numbers of full-time and part-time faculty based on nationality, 2019," accessed on February 1, 2021, http://acct2019.cc.ntu.edu.tw/acct2019e/acct6/3.pdf.

overseas – given Taiwan's somewhat precarious geopolitical situation. The United States continues to be the top destination for Taiwanese graduate students. In the 1980s, entire graduating cohorts from Taiwan's top universities would move to the USA for graduate studies, making Taiwan the top sending nation for international graduate students in the USA.[146] However, there has been a general decline in the number of Taiwanese graduate students going to the USA, with the annual numbers decreasing from 15,022 in 2000 to 9,236 in 2017.[147] This is in part because more of these students are choosing to stay at home in Taiwan. However, not all are pursuing graduate studies at home. Overall, Taiwan is also producing fewer and fewer science and engineering graduates each year, partly because of Taiwan's declining birth rate and also because a growing number of Taiwanese students see other disciplines as more attractive.[148] This decline puts greater pressure on Taiwan to make itself an attractive destination for international graduate students in the sciences as well as international faculty, especially as its much larger neighbor, China, tries to do the exact same thing.

Science and Technology in Taiwan

Taiwan's investments in academic research in the sciences started soon after the end of the Second World War. Academia Sinica (the equivalent of the National Academy of Taiwan) was reestablished in Taipei in 1949 with a focus on mathematics, philology and botany. Academia Sinica continues to be the premier research academy in the country, reporting directly to the Taiwanese president and enjoying a dedicated budget and significant autonomy in its decision-making. It comprises three divisions, one of which is focused on the life sciences. Within the Life Sciences Division, there are nine independent research institutes that focus on a range of subfields within the biological sciences, from plant biology to genomics.

---

[146] Saxenian 2002. This is no longer the case, with Taiwan now the sixth largest source country for international students in the USA.
[147] Evan Feigenbaum, "Assuring Taiwan's Innovation Future," Carnegie Endowment for International Peace, January 29, 2020, accessed on December 14, 2020, https://carnegieendowment.org/2020/01/29/assuring-taiwan-s-innovation-future-pub-80920.
[148] Feigenbaum, "Assuring Taiwan's Innovation Future."

At the end of the twentieth century, the Taiwanese government passed the Fundamental Science and Technology Act,[149] which reaffirmed its commitment to promoting scientific and technological development in order to raise the standards of both basic and applied scientific research in the country and to boost economic development. As part of the act, the government is required to formulate a National Science and Technology Development Plan once every four years to provide an overview of the current state of scientific R&D in the country and set goals for the next four years. This plan is to be developed with advice and input from Academia Sinica. Thanks to this influence asserted by Academia Sinica, basic research is supported by the government, but at levels lower than in comparable countries. Overall, Taiwan's R&D expenditure was 3.06 percent of its GDP in 2015, lower than that of South Korea (4.23 percent) and Japan (3.49 percent), but higher than that of China (2.07 percent) and the United States (2.79 percent).[150] The National Council on Science Development was also renamed the Ministry of Science and Technology in 2014 – raising its status within the government bureaucracy – and assigned the goal of promoting and funding academic research in the sciences.

Bioscience Research in Taiwan

In the 1980s, Taiwan identified biotechnology as one of its new strategic foci for future technology development, hoping to replicate its earlier success with the semiconductor industry.[151] The Development Center for Biotechnology was established in 1984 as a nonprofit organization with the principal mission of developing new drugs and creating patented technologies for use in new, innovative bioscience products. In the same year, the National Science Council began to undertake large-scale biotechnology projects, while Taiwan's first biotechnology company was established in Hsinchu Science Park to produce Hepatitis B vaccines for the world market.

---

[149] "Fundamental Science and Technology Act," Laws & Regulations Database of The Republic of China, amended June 14, 2017, https://law.moj.gov.tw/Eng/LawClass/LawAll.aspx?PCode=H0160028.
[150] Science Technology & Innovation Center 2004.
[151] Hsu et al. 2005; Chen and Huang 2004; Dodgson et al. 2008.

By the mid-1990s, biotechnology was seen as the country's next "star" industry.[152] The government began vigorously promoting research in related industries as well as encouraging overseas Taiwanese life scientists to return and contribute their research experience and expertise to the nation.[153] Similar to Singapore, the Taiwanese government invested in four interrelated areas of biotechnology: manpower development, R&D infrastructure investments, fiscal incentives to encourage private investments in R&D, and the development of government-linked research centers and companies. Thanks to this strong push from the government, biotechnology R&D in Taiwan's science parks increased rapidly from 1991.[154] In 1993, Academia Sinica established the Institute of Molecular Biology and the Institute of Medical Biotechnology.[155] Three years later in 1996, the National Health Research Institutes (NHRI) were established to spearhead biomedical research in Taiwan, along the same lines as the US National Institutes of Health (NIH).

The early 2000s also saw significant investments by the Taiwanese government in the bioscience and biotechnology sectors. Programs to accelerate the transfer of new discoveries into scalable products were established. The National Research Program for Genomic Medicine was launched in 2002, while the Experimental Center of Biotechnology was established within the Southern Taiwan Science Park in 2004. In 2007, then-Taiwanese president Chen Shui-Bian announced the goal of the biotechnology industry becoming Taiwan's next trillion-dollar industry. This was followed by the opening of the Hsinchu Biomedical Science Park in 2008.

---

[152] Wong 2005.
[153] Return migration has been a key driver of Taiwan's economic success from the 1990s onwards, and continues to be heralded as a path to the development of its biotechnology sector (Iredale and Guo 2003). However, return migration rates have fluctuated significantly over the decades (Chang 1992).
[154] STIC 2004.
[155] There are currently nine research institutes and centers within the Division of Life Sciences at Academia Sinica ("About Academia Sinica," Academia Sinica, October 26, 2016, accessed on December 14, 2020, www.sinica.edu.tw/en/articles/12).

Present and Future Challenges

By 2020, there were seven science parks spread throughout Taiwan, hosting more than 1,100 biotechnology companies involved in drug manufacturing, herbal medicines, R&D for new drugs, medical device manufacturing and implants.[156] Still, Taiwan's relatively small domestic market and Taiwanese companies' lack of international marketing experience have been identified as ongoing structural weaknesses.[157] Taiwanese investors, perhaps having grown used to the relatively short turnaround times for investments in the electronics industry, can also be impatient about the slow and unpredictable return from biotechnology investments. A vertically integrated biotechnology industry, which would allow innovations from research labs to be rapidly translated and commercialized by downstream industries, has not yet developed.

Given that it is still in its infancy, Taiwan's biotechnology industry faces significant competition from neighboring Asian states that are implementing similar strategies and trying to recruit from the same high-skilled manpower pool. Published in 2017, Taiwan's four-year National Science and Technology Plan identified a strong need to cultivate "a new generation of innovative R&D talent" and noted that Taiwan "needs to create an environment that attracts worldwide talent to further safeguard the international cooperation network for scientific talent" (MOST 2017:9). But Taiwan tends to lag in this regard with limited internationalization seen in its R&D investments.[158] Additionally, return rates for US-trained science and engineering PhDs from Taiwan have declined somewhat from their highs in the 1990s, making further internationalization of the R&D sector somewhat challenging.[159]

A further complication is Taiwan's geopolitical and economic position vis-à-vis China.[160] There has been a steady shifting of FDI from Taiwan to the Chinese mainland, which offers lower costs of

---

[156] "Why Taiwan," Biotechnology and Pharmaceutical Industries Promotion Office, MOEA, accessed on March 24, 2020, www.biopharm.org.tw/en/why-taiwan.html.
[157] Chen 2005.
[158] Feigenbaum, "Assuring Taiwan's Innovation Future."
[159] Feigenbaum, "Assuring Taiwan's Innovation Future."
[160] Many scholars have noted the importance of studying a country's scientific development alongside its internal political development and external geopolitical status (Jasanoff 2005; Ezrahi 1990; Shapin and Schaffer 1985).

production, a larger domestic market and, increasingly, an equally educated labor pool. China has also been flexing its muscles more aggressively with regard to Taiwan's status as an independent country. This will inevitably make foreign companies wary of making too big an investment in Taiwan given the risk of future instability. It is difficult to ascertain what could be the best long-term R&D strategy for Taiwan in this challenging environment.

Conclusion

This chapter has provided readers with an overview of my four Asian country case studies. The goal of this chapter was to provide sufficient background context to readers unfamiliar with the history of the scientific R&D and higher education sectors in these four countries. This chapter has also highlighted some of the shared experiences of development, modernization and investment in the life sciences, and also some of the unique challenges faced by these countries.

Detailing the scientific histories and progress of these four countries side by side allows for illuminating comparisons. In all four countries, we see the key role played by the state in setting the agenda for scientific research, investing in ambitious research projects, and reshaping existing educational and scientific infrastructures within each country to pursue long-term developmental goals. The volume of state funds being earmarked for scientific R&D in all four countries has steadily increased in recent decades, allowing researchers in these countries to be ever more ambitious. But it is not just funding, but also other dimensions of the scientific research systems in these countries – such as the technological infrastructure and research networks – that have objectively improved since the turn of the century as well as the corresponding research output.[161] In all four countries, we see Mariana Mazzucato's (2013) idea of the "entrepreneurial state" in action – investing in high-risk-high-return programs that individual corporations may avoid, building an infrastructure upon which companies can leverage their business ideas, and adopting a longer time-horizon than most publicly listed companies are able to.

But we also see significant variations within this broad approach. This chapter highlights the unique constraints that each of these four

---

[161] Marginson 2021.

Asian countries have had to work within and the areas where significant obstacles remain. The Taiwanese state must navigate an international geopolitical system in which it is increasingly isolated. Even the Singapore state must constantly demonstrate its relevance and use value to foreign investors and even scientists given the increasing mobility of economic and human capital. Despite its wealth, Singapore's tiny native-born population means that it will always be heavily dependent on foreign researchers and foreign companies in ways that the other three countries will not. The Indian state needs to revamp its higher education sector to meet its goals for a nationwide economic and scientific transition. But it is grappling with two choices: reforming the system from within versus starting from scratch with new educational institutions built outside the existing system. In many ways, India is the furthest behind among my four case countries, with a national research infrastructure that is not as advanced or as well funded as what is available in the other three, and a higher education infrastructure that continues to be beset by inefficiency and bureaucracy. And yet, it also has the youngest population out of the four case countries, with steadily growing numbers of graduates in science and engineering. Finally, China needs to address research problems that money alone cannot solve, such as improving the actual and perceived quality and validity of its scientific output while boosting the impact of its research.

All four countries have to guard against the danger of creating a culture of complacency and dependency within their scientific community, as well as the business community – who may assume that the state will or should always be there to fund their research. States also need to ensure that they are able to reap the rewards of their scientific investments and that these are not passed on to other countries that are better equipped with a stronger intellectual property or industrial infrastructure, to allow easier translation of research discoveries into businesses.

In all four countries, we also see rapid changes occurring within their higher education sector. From largely teaching-oriented institutions, the top national universities in these countries have been given the mission of becoming more research focused, and more ambitious and entrepreneurial in their long-term vision, so as to compete with established universities in the West. The competition between these universities is also heating up as they each seek to be ranked higher than their peers as world-class research universities.[162] In all four countries, we see a reliance on the return migration of Asian scientists as a way to jumpstart or accelerate this scientific and educational progress. The return of these Asian scientists is the focus of this book, and a central question I am concerned with is what these returning Asian scientists bring back with them in terms of their normative ideas of how to *do* and how to *teach* science.

These questions will be addressed in subsequent chapters as I shift from the macro perspective to the micro-level accounts of the individual scientists I interviewed. In the following chapters, readers will be able to see how these issues play out in the personal decisions of interviewees to leave their home country to seek training in the West, their decision to return to Asia (or not) and, for those who did return, their experiences conducting research in a rapidly changing Asia. In addition, the following chapters will allow the Asian-born scientists I interviewed to provide their more personal accounts of how *they* see the scientific and educational terrain in Asia shifting and how they see themselves agentically reshaping the national research landscape in their respective countries, through their migrations, teaching and research.

[162] Marginson 2021.

PART II

Circulations

3 Leaving Home, Heading West

> I knew I definitely wanted to do a PhD in neuroscience and I was interested in vertebrate neuroscience and there wasn't much in the '90s [in India]. This was 1992, 1991. I looked around the country and knew [India] was at that state that it made sense for me to go abroad to do my PhD because the kind of work I was interested in just wasn't available to me in my country.

This recollection comes from Sneha, an Indian neuroscientist who completed her PhD in the USA in the 1990s and then pursued postdoctoral work in another country, before returning to India. Sneha's retelling of her decision to leave her home country parallels the account of my oldest Chinese interviewee, Wang Wei, who was born in 1950s China and lived through the Cultural Revolution. He trained as a medical doctor in China, before joining a PhD program in Europe. He explained to me how Chinese science students of his generation had viewed the West:

> In the '80s, China started opening the door to see the Western education. Then we start to know the journals like *Lancet* or the *New England Journal of Medicine*. About *Lancet*, we will say, "This is the best of the best!" ... And then, China just opened in 1980 and then people, they go out [of China] and say, "Wow! Outside, the medical science is so good, and the equipment, knowledge and everything."

These accounts from two of my interviewees represent a key factor in the decision made by many aspiring Asian scientists in the second half of the twentieth century, to seek graduate training in Western countries. All were confronted by a dearth of training opportunities in Asia and a concurrent wealth of opportunities in the West.

This chapter introduces the Asian scientist migration system to readers, focusing on the first stage of this migration system: the initial migration to the West for training. In all four of my Asian case studies, a well-established training pathway to the West had emerged in the second half of the twentieth century because of the differences in scientific advancement levels between Asia and the West.

But Western countries' relative scientific prowess was not the only factor that drove my Asian interviewees to seek training programs in the West. This chapter lays out the multiple, overlapping motivations and mechanisms behind aspiring Asian scientists' migration to the West. Drawing on my four Asian country case studies – China, India, Taiwan and Singapore – I show how overdetermined the desire to seek training in the West was in the twentieth century. The factors that drove my interviewees to go to the West for training included the structural inadequacies of the scientific training system within their home country, government- and university-driven opportunity structures in the West and in Asia that encouraged westward student migrations, the cumulative network effects driven by earlier cohorts of Asian student migrants in the West, and the widely circulating images of specific Western countries as scientifically advanced and welcoming.

However, because of the unique nature of my interviewee sample, which spans Asian scientists who first left their home countries in the 1980s and earlier to those who left in the mid-2000s, this chapter also traces how the relative importance of each motivating factor has changed over time, and how interviewees' assessment of the relative strengths of the scientific training system in Asia has also changed. These changes speak to the rising standard of science education in various Asian countries vis-à-vis the West, and the impact this is beginning to have on the training destination decisions of future Asian scientists.

The final takeaway from this chapter is the growth in the number and variety of education and training pathways now available to aspiring Asian scientists. The dominant pattern of migration in the twentieth century – undergraduate studies in Asia followed by doctoral and postdoctoral training in the West – is no longer the only viable option for an ambitious aspiring scientist from Asia (see Figure 1.3 in Chapter 1). With the growing stature of national research universities in select Asian countries, some aspiring Asian scientists may voluntarily choose to complete their doctoral training in their home country and only move to the West for postdoctoral training.[1] Other aspiring

---

[1] Paul 2018; Paul and Long 2017.

scientists may choose to move *within* Asia for graduate training as regional training hubs begin to emerge. These increasingly acceptable alternative paths to becoming a scientist in Asia indicate how the global scientific field has changed over the last few decades, leading to a diversification of the Asian scientist migration system.[2]

Before I discuss these findings, however, I outline the general causal mechanisms behind international student migration and aspiring scientists' training-related migrations. I then organize the remainder of the chapter around the three interlinked spatial and temporal training decisions my interviewees had to make:

(1) *Whether or not* to leave their home country and seek training in the West,
(2) *When* in their career course to leave, and
(3) *Where* in the West to go for this training.

My interviews reveal that the thinking around these questions started shifting in the 2000s, as a growing number of aspiring Asian scientists in my sample began adopting alternative training pathways. I end the chapter by discussing how interviewees who returned to Asia have a new understanding of the educational and training possibilities in Asia in the present moment, influencing the advice they give to their current students – the next generation of Asian scientists – about where to train. This could potentially lead to even more changes in the training pathways of future Asian scientists, and even more complexity being added to the Asian scientist migration system.

International Student Migration

The international student migration literature largely focuses on the factors in sending and receiving countries that lead to the emergence of this international migration stream.[3] Within this body of research, much has been written about the specific case of Asian students moving to the West for tertiary degrees.[4]

In Asian origin countries, a shortage of university seats and university programs of questionable quality have been found to drive

[2] Ortiga et al. 2018.
[3] Altbach 1998; Li and Bray 2007; Hung 2010; Mazzarol and Souter 2002.
[4] Findlay 2010; Hamilton, McNeely and Perry 2012; Hung 2010; King and Raghuram 2013; Li et al. 1996; Mazzarol and Souter 2002; Raghuram 2013.

aspiring undergraduate students (and their parents) to consider an overseas college education.[5] Simultaneously, however, worries about being apart from family, a lack of proficiency with foreign languages (particularly English) and, of course, the high cost of an international education may make some Asian students reluctant to leave home.

In the destination country, universities located in global cities, and with reputations as world-class institutions, draw international students who are seeking a personal experience of what they understand as modernity and/or cosmopolitanism.[6] Universities in Western countries are increasingly relying on higher-fee-paying international students to shore up their shrinking budgets so they actively market themselves in overseas markets (particularly in Asia).[7] Destination country governments also tend to be more lenient towards the entry of international student migrants, especially in priority fields such as STEM. They may introduce special visa categories or scholarships for international students, recognizing the disproportionately large and positive impact these students can have on the host country's future economic growth.[8] In the USA, for example, all international tertiary students are automatically granted an additional one-year "Optional Practical Training" (OPT) period after they graduate, during which they can seek employment without applying for a separate work permit. Meanwhile, international students who graduate with a STEM degree are given an additional 24-month extension to the standard 12-month OPT period,[9] thus making the USA an attractive destination to international STEM students who want some overseas work experience before returning home. This policy partly helps explain why US five-year stay rates for international doctoral students in science and engineering can be over 80 percent for some Asian nationalities.[10]

---

[5] Mazzarol and Souter 2002; Li et al. 1996.
[6] Raghuram 2013; Robertson 2006.
[7] Hegarty 2014; Choudaha and Chang 2012.
[8] Milio et al. 2012; Kapur and McHale 2005.
[9] "Optional Practical Training Extension for STEM Students (STEM OPT)," U.S. Citizenship and Immigration Service, updated February 26, 2021, accessed on June 1, 2018, www.uscis.gov/working-united-states/students-and-exchange-visitors/students-and-employment/stem-opt.
[10] Finn 2014. In 2017, the five-year stay rate for Chinese and Indian doctoral graduates in science and engineering was 83 percent, compared to the overall stay rate of 71 percent (Amy Burke, "Immigration and the S&E Workforce," National Science Foundation, September 26, 2019, table 3-22, accessed on

Strategically minded students who are not interested in settling down overseas may still seek an overseas degree if they believe that this type of qualification will improve their chances for upward mobility upon their return home.[11] (In Chapter 2, I discussed how students from China, India and Taiwan all see an overseas degree as giving them an edge over their domestically trained competitors.) Other students may hope to emigrate permanently and they may view a degree from their preferred destination country as a first step to securing an overseas job in, and long-term visa to, that country.[12] Their student visa effectively becomes a gateway to permanent residence.

Waters and Brooks (2010), however, push back against the idea of the "strategic international student migrant." The authors' argument that not all migration decision-making follows a conscious, rational choice model is an important reminder that the migration and destination decision-making process is a subjective one, framed by each individual migrant's socially constructed dreams for themselves and their mental maps of the world.[13] As I show later in this chapter, a migrant's decision-making may be simultaneously strategic *and* subjective. Prospective doctoral student migrants can be as influenced by affect and subjective aspiration, as by an objective evaluation of the costs and benefits of different choices. Like all other migrants, they too may make migration and destination choices using imperfect information and driven by implicit/explicit preferences and biases.

But there are ways in which international *graduate* student migration is different from other forms of international student migration. The international migration of doctoral students in the sciences, in particular, sits at the intersection of student and scientist migrations, as it is often at this point that individuals see themselves as embarking upon a career course that will end with them becoming practicing scientists (see Figure 1.1). Like other international students, most international doctoral students enter their host country on student visas and work towards a degree. In this sense, they are students. However, doctoral students tend to be significantly older than most

---

June 26, 2021, https://ncses.nsf.gov/pubs/nsb20198/immigration-and-the-s-e-workforce.

[11] Xiang and Shen 2009; Hung 2010; Zweig et al. 2004; Chang 1992.
[12] Rosenzweig 2007; Mazzarol and Souter 2002; Li et al. 1996; Hamilton, McNeely and Perry 2012.
[13] Paul 2011, 2017.

undergraduate migrants and are less likely to be influenced by their parents' aspirations for them compared with the preferences of their peers and professors.[14] Hamilton, McNeely and Perry (2012) point to the importance of *overseas* social networks of conational professors and peers in the cumulative migration of international doctoral students from the same country (and even the same undergraduate institution) to particular overseas destinations, and I find similar influences among my interviewees.

Training, however, does not stop after doctoral studies, and neither do scientists' migrations. As mentioned in Chapter 1, aspiring academic scientists increasingly have to undertake postdoctoral training before securing a faculty position, especially at top research universities and institutes around the world. So it is worthwhile considering the factors that influence international migration decisions made during the postdoctoral training phase. Furthermore, the postdoctoral destination decision has been found to be more impactful than the doctoral destination decision when it comes to the formation of research networks and cross-national collaborations.[15]

In 1979, the number of *foreign* postdoctoral trainees in science, engineering and health fields in the USA was 6,065. By 2015, this number had increased almost six times to 35,135 foreign trainees, amounting to 55 percent of the total number of postdoctoral trainees in the country.[16] A significant proportion of these foreign postdoctoral appointees completed their doctoral degree elsewhere before coming to the USA,[17] with one estimate putting the number at more than 50 - percent.[18] In fact, the United States is the single largest destination for overseas postdoctoral training.[19] In addition to Asian doctoral holders,

---

[14] Carlson 2013; Vertovec 2003; Szelényi 2006; Xiang and Shen 2009.
[15] Woolley et al. 2008.
[16] "Survey of Graduate Students and Postdoctorates in Science and Engineering, Fall 2015," National Science Foundation, February 2017, table 31, accessed on February 1, 2021, https://ncsesdata.nsf.gov/datatables/gradpostdoc/2015/html/GSS2015_DST_31.html.
[17] "Survey of Graduate Students and Postdoctorates," table 50, accessed on February 1, 2021, https://ncsesdata.nsf.gov/gradpostdoc/2015/html/GSS2015_DST_50.html. This NSF survey is frequently unable to capture data on the country where postdoctoral appointees completed their doctoral degrees, so these estimates are based on only 65 percent of all postdoctoral appointees for whom country data is available.
[18] Stephan and Ma 2005.
[19] Franzoni et al. 2012.

European and Australian doctorate recipients also often choose to pursue their postdoctoral training in the USA.[20]

Melin and Janson (2006) explain the flows of European PhD recipients to the USA for postdoctoral training by positing the existence of an American "halo effect." They write that American universities' reputation for being highly selective draws European PhD recipients to seek American postdoctoral training in order to gain the scientific cultural capital of having been selected by a US institution. There is also entrenched institutional support within European scientific circles for a *lateral* form of brain circulation to further a young scientist's intellectual development.[21] The conviction in Europe is that international mobility exposes postdoctoral trainees to different ways of approaching scientific questions regardless of the relative quality of scientific research in any given country (though this is still largely limited to circulations within Europe and the USA). This mindset encourages international mobility among junior European researchers who may complete their doctoral education in Europe (though not necessarily in their own country) and then seek postdoctoral training in the USA, where they can acquire not only scientific human capital (in the form of advanced training in their subfield) but also location-specific scientific cultural capital (as a result of gaining the status of being "American-trained"). As I show later, this particular set of motivations parallels what is emerging within the Asian scientist migration system as well.

But the upshot of recent improvements in the doctoral training scene in Asia, alongside the rise in the global research profiles of select Asian national universities (discussed in Chapter 2), is that aspiring Asian scientists now have local, regional and global training pathways to choose between (see Figure 1.3), while in the past they may have had only one respectable doctoral and postdoctoral training option: going to the West (see Figure 1.2). In the remainder of this chapter, I share my interviewees' accounts of their decisions to seek training in the West. I draw on the wide range of their initial departure dates from the 1960s to the early 2000s, to highlight the increasingly acceptable training options now available to aspiring Asian scientists in Asia itself.

---

[20] Moguérou 2005; Melin 2004; Pellens 2012; Turpin et al. 2010; NISTADS 2009.
[21] Ackers 2004; Commission of the European Communities 2001.

Deciding Whether to Leave

Among my interviewees, the scientist with the earliest doctoral degree was a Malaysian who began his graduate studies in the West in the 1960s. However, the majority (76 percent) of the 119 scientists I interviewed left Asia in the 1990s and 2000s (see Table 1.4 in Chapter 1). This wide range in the decade of departure from Asia allows for comparisons across three broad time periods (the 1980s and earlier, the 1990s, and the 2000s)[22] regarding the factors that led interviewees to pursue training in the West. In the following sections, I detail the four factors shared across my interviewees that explain the emergence of this transnational training pathway connecting Asian and Western countries:

(1) The superior standing of Western countries/universities versus their Asian counterparts in the global scientific field,
(2) The transnational migration and training opportunity structures that set the stage for interviewees to leave Asia and travel to the West for graduate and postgraduate training,
(3) The norm of heading West for graduate training, which was pervasive in the top national universities in all four Asian country case studies and was fostered by networks of earlier migrating cohorts of students and professors, and
(4) The positive media-based images of Western universities and the practice of science in Western countries.

Throughout, I also highlight how each factor's relative importance has shifted over these three time periods, thus setting the stage for the recent changes in the Asian scientist migration system.

The Relative Position of the West in the Global Scientific Field

Before the 1980s and all the way into the 1990s, there was widespread recognition that scientific expertise in the biological sciences and technological superiority was overwhelmingly concentrated in the West and, within the West, in the USA. As indicated in the accounts

---

[22] I collapse the interviewees who left in the 1980s or earlier into a single category as the number of interviewees who departed in the 1960s and 1970s is not large, and the motivating factors they identified were not that different from those identified by interviewees who left in the 1980s.

of Sneha and Wang Wei that started this chapter, interviewees who sought scientific training during this time period overwhelmingly saw countries in the West as offering superior training. The most critical factor was often the lack of any domestic doctoral program that focused on their specific research interest. One Chinese bioscientist recalled that, in the early 1980s, when he was considering graduate studies in biology after finishing his undergraduate degree at the top university in his province, he faced the dilemma of not finding any professors able to supervise him because there were so few faculty with PhDs who were permitted to take on doctoral students.[23]

There was also an acknowledgement that, within the West, the USA offered the best training. Wai Ming, who trained in the West in the 1960s, shared that when he was considering where to pursue doctoral training, his master's degree supervisor (who had trained in Canada) advised him to go to the USA instead: "He told me to go to the US because the US is still the leader in science."

This sense of the relative standing of different countries' scientific research systems began shifting in the 2000s. By that time, there was a growing recognition among my interviewees that the research infrastructures and technical training available in the very top Asian universities was improving and drawing apace with what was available in all but the very top universities in the West – due in part to these Asian universities' heavy investments in improving their scientific research systems. In many ways, gaining admission into and completing a doctoral degree at a top Asian university, such as Tsinghua University in China or the Indian Institute of Science, gave credibility to an individual's capabilities as a junior researcher, ensuring that they would secure prestigious postdoctoral fellowships in the West. Anmei, a Chinese bioscientist who studied medicine in China before moving to the USA, shared that, starting in the mid-to-late 2000s, graduating from a place like Tsinghua or Peking University in China exerted a powerful signaling effect:

> I think the education in China was wonderful. I have relatives, friends, who always ask me if they should send their kids here [to the USA], when they are in China. And I always tell them, "If you think your kids are very competitive, one way to prove it is to go to the top university in China."

[23] A similar situation has been found to have existed in Taiwan (Chang 1992).

Another interviewee, Priyanka, who completed her undergraduate degree in India at a regional university and then was accepted into an integrated masters and PhD program at one of India's top science universities in the late 1990s, had no regrets about *not* going overseas for her PhD. As she put it, she was "accepted into the best program for very gifted kids in [India]" and so had no reason to decline the offer. Likewise, another Indian scientist, Shiva, opted to stay in India for his PhD in the 2000s because he was awarded a prestigious national doctoral fellowship from one of India's premier research institutes:

> So my initial schooling was in [India] and then I got a ... fellowship. That's a national fellowship. I think those days it was probably only fifty seats per year [for the fellowship] on all science fields together, and biology like only ten or so seats. It was quite good to get this fellowship, and the good thing is this gave me five years of fellowship.

As I elaborate later in this chapter, these changing views of the quality of scientific training in certain pockets of Asia vis-à-vis places in the West influence how aspiring Asian scientists think about the need to go overseas for their training.

Transnational Migration and Training Opportunity Structures

For those interviewees who pursued their graduate training in the West in the 1980s and 1990s, various macro-level factors in both their origin and destination countries encouraged their decision to pursue their doctoral studies overseas.

As I mentioned in Chapter 2, several Asian governments established scholarship programs to send the brightest students in their country to the West for doctoral training, with the condition that these scientists-in-training would subsequently return home and bring their overseas-acquired scientific expertise back with them. The Chinese-US Biochemistry Examination and Application (CUSBEA) was founded in 1982, following the 1979 reestablishment of diplomatic relations between China and the USA, to enable a select group of talented students from China to pursue doctoral training in biology and chemistry in the USA on full scholarships. Between 1982 and 1989, 422 Chinese students traveled to the USA for doctoral studies in biochemistry or structural biology under the auspices of CUSBEA (Chang

2009). Likewise, the Singapore and Taiwan governments funded the overseas PhDs of their brightest local students during the 1980s and 1990s, with the requirement that they would subsequently return home to teach in the tertiary education sector.

But the higher education policy framework in their home countries was not the only factor that encouraged aspiring Asian scientists to look westward. The immigration policy structures and degree-funding arrangements of Western countries and Western universities made entering as a doctoral student both cheaper and easier to accomplish, compared to entry at earlier points in an aspiring scientist's education trajectory. Up until the 2000s, the best time for aspiring Asian scientists to move to the West was the PhD moment. Suresh, an Indian-born, US-educated researcher, explained that he had seen his doctoral migration to the USA in the early 1990s as his "one chance" to gain relatively easy entry into the West:

> I felt very guilty when I actually left. But I left because I felt that this was this one window of opportunity for me that would close if I did not avail of it. This is my one chance. And I will never know what I am missing and I will always wonder about it [if I don't take it].

There was a widespread sense in the 1990s and earlier that the only affordable opportunity that aspiring Asian scientists had to move to the West was at the PhD moment. Financial aid or scholarships to attend undergraduate programs in the West were rarely available. Likewise, master's programs typically expected international students to source their own funds to cover the high tuition fees. Only a handful of interviewees had access to enough family wealth to travel to the West on a one- or two-year master's program, which they paid for out-of-pocket, before progressing to a university-funded PhD program. For most, the fees and funding structures of Western tertiary education programs, and the regulatory regime governing student visas to Western countries, privileged the doctoral moment as the ideal point in their career course for a westward migration. Leaving later at the postdoctoral moment was somewhat challenging in the twentieth century because the quality of Asian doctoral degrees was not always legible to Western universities, leading to science departments in the West setting a very high bar for postdoctoral applicants direct from Asia.

Norms and Networks of Leaving

Even as macrostructural factors influenced the westward migration of aspiring Asian bioscientists in the 1980s and 1990s, a culture of migration to the West was also pervasive within the top science and engineering departments across Asia.[24] The normalization of this migration pattern was reinforced by university students' peers, professors and earlier student cohorts who had already migrated to the West.

Suresh, introduced earlier, described to me the culture of migration that prevailed within the particular Indian Institute of Technology (IIT) campus where he pursued his undergraduate degree:

> Where I did engineering [in an IIT in the early 1990s], everyone would leave the country. The entire batch would just get up and leave, right? And do a masters or a PhD somewhere in the US. That was kind of the default.

This culture of westward migration for graduate study dissuaded most aspiring scientists in the top science and engineering departments in Asia from considering domestic options for their doctoral education. A researcher from India told me, "It's like a train. It's like you want to be on that train." Another scientist described the process as a "pipeline" and noted the strong influence of earlier cohorts within his university: "I certainly think that one of the influences was these undergraduates ... who had gone before us."

Yue Wang, a US-trained Chinese scientist, recollected just how prevalent this desire to go overseas was at his Chinese university, even though it was renowned within China for its science education programs. He calculated that more than half of his graduating cohort attempted to go overseas for graduate studies, and described how this aspiration became normalized among the top students at his university:

> The university I went to had a very good tradition of many people wanting to go abroad to get degrees. So when you first enter the university ... there would be many people selling their old books ... on vocabulary of different levels of English. And university-wide, many, many people, probably more than 50 percent of the people actually went to take the GRE as part of the

---

[24] Such cultures of migration have been observed in many different contexts. See Kandel and Massey 2002; Ali 2007; Paul 2017.

experience. So from the very beginning, I had a conscious decision that I wanted to do this as well.

This normative culture continued into the 2000s, though like their European counterparts, interviewees who left during this time period increasingly stressed the importance of lateral brain circulations as a means for encountering new ways of doing science. More of these interviewees also chose to pursue their PhDs in Asia, postponing their departure to the West until their postdoctoral training moment. Hanna, an Indian bioscientist who received her PhD in India before going to Europe in the early 2000s on a postdoctoral fellowship, explained the importance of overseas training in a manner that contrasted strongly with the explanations from earlier interviewees:

> Because the world is getting more and more connected. If you don't leave home and go to a place which is totally different, (a) you will not understand how to make yourself relevant globally, and (b) you won't know what your country is all about. So I think it's really important. Everybody should go and get exposure. Go to some culture anywhere outside.

Scientists like Hanna, who left their Asian home countries in the 2000s, were less likely to frame their westward migrations as making up for a technological or scientific "lack" in Asia. Instead, the reasons for moving to the West were increasingly characterized in terms of coming into contact with key figures in their scientific subfield who were still largely concentrated in the West. As Malcolm, a scientist who studied and then taught at an Ivy League university for several years before returning to Taiwan, explained:

> It's the connection. It's the networking. You'll get to know this big name and you will have worked with this [big person]. This is really about connecting with the key leaders in a certain field. ... In my field, if I mentioned the name of my mentor [in the USA], everyone would know him ... Of course, it's content that I learned from him. Of course, that's invaluable. But, when I say that I have been trained by so-and-so and I have worked in so-and-so's lab, that's also helpful. ... [Now,] I don't want to give you the impression that the connection is all you need. But I do want to say that even if you have the best science, if you have none of this networking, it will limit someone's development.

Interviewees saw these network connections as a form of scientific social capital that was location-dependent in its accrual.[25] Such connections mattered because this was how job information was passed along, referrals were made, new joint projects initiated, grants secured, new findings circulated and international reputations built. The scientists I interviewed were very aware of the "rules of the game" in the global scientific field, as I highlighted in Chapter 1. "Networking is about resources. If you have good networking, definitely you have better resources," was how another scientist put it to me.

Asian-based networks were important too but in other ways. For individuals who were certain that they would return to Asia, home-country networks could be essential to their career success with grant applications and promotions in the home country. And so it was often useful to also nurture those local connections. But there was a sense, even during my interviews in the mid-2010s, that most star scientists[26] in the biological sciences were still located in the West, and particularly in the USA. Time spent in these countries, ideally under the tutelage of or in close proximity with one of these star scientists, was deemed essential for a junior Asian scientist's career. Thus, scientific networking (rather than gaining access to superior research infrastructures) became an increasingly prevalent justification for westward migration in the 2000s and beyond.

Media-Driven Imaginaries of Western Science

Anthropologist Arjun Appadurai writes about "mediascapes" as "image-centered, narrative-based accounts of strips of reality ... that constitute narratives of the 'other' and proto-narratives of possible lives" (1990:299). He describes how these "fantasies" can kickstart certain aspirations, which in turn lead to particular decisions that impact an individual's future life outcomes. Applying this concept to scientist migration reminds us that it was images of and narratives about Western science, not firsthand experience, that inspired interviewees' aspirations to move westward. The vast majority of them had

---

[25] It is still an open question how location-independent the maintenance and utilization of these scientific network connections are, but certainly the accrual of these connections required individuals to be physically present where these networks were concentrated.

[26] Zucker and Darby 1996.

never visited the West beforehand. Their graduate training decisions were based on the prominent position the West enjoyed in their mental maps of the world, which they had constructed from the secondhand information gleaned from various mass communication channels as well as their social contacts.[27]

Prior to the 1989 birth of the World Wide Web, other media technologies (radio, print magazines, newspapers, films, television, etc.) were used by interviewees to construct their geographical imaginaries of potential universities and countries to apply to. Jay, a Chinese scientist, spoke of learning about the USA through illicit American radio channels before China reestablished diplomatic relations with the USA:

> Back then, we knew nothing and the only channel we knew anything about outside China is the *Voice of America* and that's how we are learning English a little bit. You know, that's actually against the law! [*Jay laughs*] But everybody was doing it. ... At that time, we thought we knew a little bit about Americans. But for the rest of other countries, we have no knowledge. You don't even know what does that look like. At least we knew a little bit about the US. Of course, later on, when you physically got there, you realized that this is different from what I imagined.

An Indian scientist, Vivek, who also left home in the early 1990s, spoke of gaining an impression of American culture through nonfiction books and films:

> We don't have the web then, so it must have been books. Certainly through my father, I read a lot of stuff, you know, nonfiction things. I started thinking of the US as this place where things could happen and things were flexible. I was right, as it turned out.

In both Jay and Vivek's cases, they openly acknowledged that they had left for the West with not much more than a hunch about what studying science and living in the USA was going to be like. In Vivek's case, he

---

[27] In this regard, they are no different from most other international migrants, who also make destination decisions by relying on subjective mental maps of the world (Paul 2017, 2011).

felt vindicated after arriving in the USA, while in Jay's case, he realized that his ideas about America had been quite off the mark.

But by the late 1990s and into the 2000s, as internet access began to spread across Asian cities, universities became hubs of this new communication technology, allowing tertiary students to communicate and exchange information with each other across great distances. Among interviewees who left their home countries during this time period, many spoke of the importance of the internet in revolutionizing how they gathered information about prospective universities in the West. Information sharing networks began to develop on online discussion boards, which connected students from different cohorts and universities within a single Asian country. In late 1990s and 2000s China, a Bulletin Board System (BBS) – an online, text-only discussion forum – was started by Chinese students to post questions and receive answers about the pros and cons of different Western countries and universities, the process of applying to universities outside the country, the minimum GPA score you would need to be guaranteed admission into a particular university, whether you were likely to be offered a scholarship, and so on. Yue Wang, who received his PhD from an Ivy League university before returning to China, described BBS as a "community" of Chinese students spread across the entire country, linked together by their common endeavor of seeking Western training. He explained how he had used BBS in conjunction with general internet resources to decide which universities to apply to:

> You had a feeling after reading the BBS. You find out that some universities are friendly to Chinese students and there have been previous admissions [from China] to the programs. And you also go to the website [of the department] to see the student profiles. And you can see how friendly they would be to international students. So a mix of randomness, what my peer students do, and also my reading and browsing the web – all helped.

Such accounts from interviewees highlight how important the internet was (and continues to be) in showing university students in Asia their options inside and outside the country, clarifying to them how to access these opportunities, and enabling them to dream of an overseas education in a more informed manner. But these information resources were not evenly distributed across and within my four Asian case countries, and this had an impact on the timing of different individuals' westward migrations.

Deciding When to Leave

Deciding to leave for the West was only the first of multiple training-related decisions that interviewees had to make on their journey to becoming scientists. They also needed to decide at *what stage in their training* they should go to the West: whether at the undergraduate, master's, doctoral or postdoctoral training moment. As Table 1.4 in Chapter 1 showed, the answer to this question changed over time. Among the factors that shaped the changing responses to this question were:

(1) The time-variant costs of funding different training stages,
(2) The timing of interviewees' exposure to advice or norms about going West, and
(3) Improvements over time in the quality of training available to them in Asia.

These factors were themselves influenced by the changing socioeconomic context in each of my four Asian case countries and their governments' increasing investment in scientific R&D. At a more micro level, interviewees' family and personal characteristics – such as whether or not they grew up in a less developed part of their country, and their parents' educational and occupational background – also influenced the timing of emigration.

The Varying Cost of Funding Different Training Stages

The timing of interviewees' first departure to the West was heavily influenced by the costs involved and interviewees' individual and familial ability to cover these costs. For degrees lower than a PhD, most international students paid their tuition fees out-of-pocket. Only seven of my Asian interviewees traveled to the West for their undergraduate education, and all of them went to the US. They either came from well-off families who could afford the high tuition fees associated with an American undergraduate education, or they were awarded government scholarships that sponsored both their undergraduate and graduate education in the West.[28]

---

[28] Chang (1992) notes that most Taiwanese graduate students who went overseas in 1979 came from relatively privileged backgrounds. Only 11 percent of the students Chang surveyed came from poor families.

One such interviewee, Shawn, was a Singaporean whose parents sent him to the USA for his undergraduate education in the 1990s. Both of his parents held tertiary qualifications and his father earned a high enough salary in Singapore such that the family could afford to pay for Shawn's four-year American degree. Likewise, the father of another Singaporean scientist, Mary, worked for a multinational corporation. During her father's postings to various countries, Mary studied in different international schools, and she was certain she would eventually study in the USA for her undergraduate education. Like Shawn and Mary, the handful of interviewees who were able to pay for an undergraduate education in the West came from Japan, Taiwan and Singapore, where their families were solidly upper-middle-class in a postindustrializing society.

Going to the West at the master's level for a one- or two-year degree was slightly more popular, because it involved a shorter time frame than an undergraduate degree. This approach was sometimes adopted by interviewees who had not known anything about PhD programs or academia as a career option when they were still in Asia, or who wanted to test the waters in the West before committing to a longer doctoral training program. Santhu, an interviewee from India, went to the USA for a master's program, which he was able to fund because he had worked at a consulting company for a couple of years in the 1990s after completing his undergraduate degree. At the time, Santhu had no idea what a PhD entailed. He told me that he had not received any prior advice – whether from his parents or his former teachers – about the possibilities of an academic career for himself:

> I didn't think of myself as one of those people who can do PhD research. We did not necessarily know what that really meant. But we understood that if you've got a master's, it was an opportunity.

After beginning his master's in the USA, Santhu started interacting with Indian doctoral students at his American university. They inspired him to pursue doctoral studies as well. Overall, however, the vast majority of interviewees only left for the West at either the doctoral or postdoctoral stage because both of these training stages tended to be fully funded.

The Timing of Exposure to Westward Advice/Norms

Among my interviewees who completed their PhDs in the 1980s and 1990s, the handful who did so in their home country in Asia, rather than in the West, reported that a lack of advice from professors and/or peers about the "right" time to go West had delayed their departures. But how was that possible when, earlier in this chapter, I noted the widespread culture of westward migration that pervaded the top national universities in my country case studies? The answer lies with the class and geographical differences within my interviewee sample.

Interviewees with parents who lacked graduate or postgraduate qualifications and/or lived away from metropolitan centers in their home country often attended mid-tier undergraduate institutions in their respective country. In that pre-Internet era, it was difficult for them to be exposed to information and advice about possible westward training pathways. These interviewees tended to take one more training step in Asia before making the move to the West, effectively delaying their departure to the postdoctoral moment.[29] This was particularly true for interviewees in my two larger country case studies – India and China – which had a much larger rural population.

One such interviewee who completed his PhD in India before going to the USA was Ram, who came from a small-business family and whose parents lacked much formal education. His father had wanted him to become a medical doctor because that was a well-trodden path in India for high-achieving students. But after Ram chose to pursue basic science rather than medicine, no one in his family knew how to advise him:

> There was no discouragement, but there was never any encouragement to go out [of India] either. At that point of time, we lived in a very small town and we being a business family, nobody has done a PhD in my family. So for our family, just the fact that someone was doing a PhD itself was a good thing. And the fact that I did my masters from [an Indian university] which is very far from [my hometown], that itself was the first big step. You know, I was leaving the home and I had never been away from

---

[29] The stepwise nature of these capital-constrained interviewees' migrations to the West finds echoes in the research conducted by Judith Zijlstra (2020) on the stepwise migrations of Iranian university students who study first in Turkey before seeking further training in Western universities.

my parents. My parents would have been very happy if I had done my PhD in the same place [in India] where I did my masters.

Ram was introduced to the idea of an overseas education after beginning a PhD program at a respected institution in northern India in the 1990s. Only after that did he leave India for postdoctoral training in the USA, an experience he found to be revelatory but also coming later in his career course than he would have liked.

Aditya also completed his doctoral training in India. After completing his undergraduate and master's degrees at a regional university in the south of India in the late 1970s and early 1980s, Aditya secured a fellowship to pursue his PhD at a national university in the north of the country. The idea of moving to the West began to develop in his mind during his doctoral years. Aditya reasoned that he had not applied to the West earlier in his educational journey because the internet was not yet widely available in India at that time. As a result, he did not have easy access to information about alternative destination options outside the country:

> I can reason it now as [the fact] that we didn't have internet. That means the only information that was available was actually, ... we only had *Nature* and *Science*, two journals that could carry advertisements for graduate admission. But by the time they came to the library of [my regional university], it's about three months old. So the position advertised failed.

For this reason, Aditya only applied to PhD programs within India (and one in Singapore), which represented the extent of his knowledge about available doctoral programs. When he was offered a doctoral fellowship to study in India, he accepted without thinking, leaving the question of overseas training for a later point in his career.

The Improving Quality of Scientific Training in Asia

By the mid-2000s, however, an alternative justification for staying in Asia until the postdoctoral moment began to surface. A handful of interviewees who left Asia during this time period spoke of intentionally postponing their first departure for the West until after they had completed their doctoral training. As the quality of instruction at the doctoral level in select Asian universities improved, alternative pathways began to emerge for aspiring Asian scientists. Increasingly, they had the option of completing their doctoral education in their home country (or *another* Asian country), as Table 1.4 shows with its increasing proportion of interviewees only leaving at the postdoctoral moment.

In such cases, the decision to head West at the postdoctoral moment was not because of a dearth of early information or advice about the value of Western training. Instead, by the late 2000s, there was growing strategizing over the timing of interviewees' first westward migration. Aspiring Asian scientists during that time period were beginning to weigh their options differently as compared to earlier cohorts of interviewees. Rather than comparing between countries or world regions, they considered the relative merits of the specific research universities/institutes where they had been accepted at home, universities in other countries in Asia, and also universities in the West. One Japanese scientist, Hiro, who completed his undergraduate, masters and PhD studies at the same Japanese university explained that, given his acceptance into what was widely considered the top science university in his country, he felt that he was already going to be exposed to the best scientific training he could find. After his PhD, Hiro secured a postdoctoral fellowship at Harvard University. He was convinced that his ability to secure a fellowship at Harvard was due to the strength of the work he completed at his doctoral university in Japan.

Other interviewees who first went to the West for postdoctoral training in the 2000s also expressed high levels of satisfaction with the training they had received in their doctoral institution in Asia. While only 10 percent (just two) of interviewees from the "1980s and earlier" cohort opted to leave for the West at the postdoctoral moment, 30 percent of interviewees in the "2000s" cohort first left for the West at the postdoctoral training moment (see Table 1.4). This shift is an indicator of the rising standing of select Asian countries' bioscience research sectors and top Asian research universities since the start of the twenty-first century.[30] Shiva only left India in the early 2000s and was adamant that the doctoral training he received in his home country was comparable to what was available in the top universities in the UK where he received postdoctoral training:

---

[30] Marginson 2021.

> I would always say, even when I was in [the UK], that my PhD in [my Indian university] was better than a PhD [from my UK university]. I'm absolutely sure about that. ... When I came out [of India] after my PhD, I was probably as good as any trained postdoc. Not just on the lab work but also any administrative work. I could write grants for my boss, I could go and present his grant when there is a meeting, take care of guests, train a lot of people. Even as a PhD student [in India], I already was taking care of other PhD students. That was the norm in our lab. There are no postdocs because those who did PhD, they all went abroad to do postdoc. So the senior-most [PhDs], then they had to take care of other trainees. And all that is just really like a complete training.

Shiva freely admitted that his university did not have the same level of resources as many Western universities, but he described a dedication to science that he subsequently never experienced during his postdoctoral training in continental Europe and the UK:

> [My Indian doctoral university] is in a semi-village, so this is really remote. Sometimes [there was] no water, no drinking water. You might have a canteen but you may not have food for students. You got to cook yourself all those things. It's a completely random place, but as soon as you step into the lab, it's totally a different place. We all students used to be around till night because daytime, we do the other things like ordering [reagents], bringing [specimens], putting them into the freezer, organizing the lab, taking care of trainees. Then when the evening progresses, that's when we do our work. It used to be [that] our boss stayed till twelve midnight everyday. His wife worked in the lab too, so she also stayed till nine p.m. His daughter goes with him; she does her homework in the lab. So that was how it was! It was quite an amazing place! And not just his lab; there were a couple of other labs that time, and they were all in their peak.

Shiva was convinced that the papers his doctoral lab in India produced were as good as those from a Western lab in his subfield. In his mind, the only thing that his Indian lab lacked was high international status, which is partly why he felt it was necessary to go to the West for postdoctoral training so that he could acquire that essential scientific cultural and social capital.

What is also revealing is the decision of some interviewees in the 2000s to seek training in another Asian country (other than their Asian country of birth) before moving to the West. Eleven interviewees adopted this approach, and their most common Asian destinations were Japan, Hong Kong and Singapore, mainly for a PhD before heading to the West for further training. Such choices speak to the rise of particular Asian countries/cities/universities as global or regional hubs for research excellence as well as potential stepping stones for aspiring Asian scientists, who employ these Asian training destinations to burnish their research credentials before attempting to go West.[31]

Deciding Where to Go

The final question that interviewees had to consider was *where* in the West to go for their training. The vast majority went first to the US. Even interviewees from former British colonies, such as Singapore and India, expressed a distinct preference for the USA, demonstrating how the UK has been displaced from its earlier centrality in the global scientific field.[32] As Table 3.1 shows, the USA consistently attracted over 80 percent of my interviewees across the three time periods I studied. The UK was a distant second, attracting only 7 percent of interviewees on average.

The popularity of the USA as a training destination was partly a case of familiarity bias on the part of my interviewees. As one Taiwanese scientist told me, "I think the USA is the country which you are really familiar with. If I know better other European countries, probably I will go [there]. But that time, I think that the USA was

---

[31] I have written extensively about stepwise international labor migration as a form of multinational migration, involving capital-constrained individuals using stepping-stone countries, which are easier and cheaper to access, as launchpads for subsequent migrations to preferred but harder to access countries higher up their destination hierarchy (Paul 2017; Paul and Yeoh 2020). However, most of this work has been in the context of low-wage migrant domestic workers. But there are clear parallels in the migration decision-making of Asian migrant domestic workers and Asian scientists, even as they are far removed from each other in terms of their class status, training and life outcomes.

[32] To learn more about the UK's centrality in earlier systems of East–West brain circulation, read Pietsch (2013).

Table 3.1 *Number of interviewees, by Western country and decade of first arrival*

Decade of first arrival	Number of interviewees (%)	Western country of first arrival		
		US (%)	UK (%)	All others
1980s & earlier	28 (24%)	23 (82%)	3 (11%)	2 (7%)
1990s	44 (37%)	36 (82%)	2 (5%)	6 (14%)
2000s	47 (39%)	39 (83%)	3 (6%)	5 (11%)
Total	119	98 (82%)	8 (7%)	13 (11%)

probably the only country I was familiar with." The term "choice architecture," coined by behavioral economists Richard Thaler and Cass Sunstein (2008), is a useful way of envisioning how interviewees' location decisions were influenced by the ways in which the USA was often presented to them as the default choice for advanced training. They had heard of American universities like Harvard, Stanford and Yale, and read the work of famous American bioscientists who taught at these and other American universities. And so they aspired to study in the USA, which they saw as being situated at the core of the global scientific field.

Often, their own professors in Asia had trained in the USA as well and these professors encouraged them to apply to the same US universities where they had studied or where they had research collaborators. In this manner, network effects influenced the specific destination decision about *where* to go in the West, as much as they influenced the migration decision about whether or not to leave home and go to the West.[33] Yong Kai, from Taiwan, explained how the head of department at his Taiwanese undergraduate university encouraged him to apply to the same research institute in the western US where the department head had trained, and so he did. Likewise, Indian bioscientist Sushmita applied to a top American public university because most of the professors in her Indian undergraduate university had studied and worked there before returning to India, and all of them told her that it was the only "worthwhile" American university to apply to.

[33] Szelényi 2006; Hamilton, McNeely, and Perry 2012.

Deciding Where to Go 107

What had interviewees known of universities in Western countries other than the USA? It was fascinating listening to interviewees offer their past impressions of different countries in the West, reflecting their sense (from the time they were students still in Asia) of how each of these Western countries compared in terms of the quality of their graduate training, openness to foreigners (and especially nonwhite foreigners), and the quality of their scientific research output. Even as they knew few specifics about many Western countries, it was clear that they did not see the West as a monolithic block of countries. In the next chapter, I will return to these variations in scientific cultures that exist across Western countries.

A key question for interviewees was whether or not a university or a country was welcoming to foreign students – in terms of graduate admissions decisions as well as its general societal openness to foreigners. This issue was important to interviewees and it drove many of them away from Europe. Others associated European countries with a more traditionalist attitude while the USA was seen as more dynamic. Sahil, an Indian neuroscientist who pursued a PhD in the USA, explained that he did not consider the UK as a potential training destination because of how he associated it with a more "conservative mindset." He acknowledged that he did not have much evidence to support his opinion but he stuck by it during his interview:

> I think I never considered the UK for even an instant. One reason was there was a professor [in my Indian undergraduate institution] who was from the UK and I did not respect that person very much. It's such a terrible thing – sample of one, right? But you know, he was not a good advertisement for the UK. Beyond that, I felt like there was a certain stodginess and conventional or conservative mindset that I associated with the UK, which was very different from what I associated with the US. ... The US just seemed like a land of opportunity. I think, even now, I would say that that was probably the right call. That was certainly true for me. I think there is a certain freedom here that I feel somehow communicates itself to people and, even back in the 90s, that existed.

The relative openness of the US to foreigners manifested itself in the fact that many scientists in the USA are first- or second-generation immigrants.[34] Interviewees talked about the US as a magnet drawing the best scientific talent from around the world. Sumanta spoke of

---

[34] Lan et al. 2015; Grubel and Scott 1966; Libaers 2007.

reading field-defining work by non-American scientists, and then realizing that they were all based in American universities, cementing his sense that the USA was where he would be most welcome:

> The people that interested me, all happened to be in the US . . . . One of them was an Australian professor who went to the US. Another one was Dutch and she also had a job in a US university. So these were not like US people, but they were at US universities.

Given the centrality of the USA in the global scientific field, what then influenced the small number of my interviewees who moved to a Western country other than the US? Often, the decision was driven by scholarship stipulations as well as peculiarities within their social networks that skewed them towards a particular non-US country. Pei-Shan, a Taiwanese bioscientist, shared that she had received a government scholarship that funded only three years of doctoral study, and given this shorter time period, she decided to go to the UK rather than the USA, since American PhDs tended to take much longer to complete. Pei-Shan applied to the only two universities she knew of in the UK – Oxford and Cambridge – without conducting any further research into these universities' programs. When both universities offered her a place, Pei-Shan said that, "At that time, we [didn't] know how to choose. Both [were] fine." Pei-Shan decided to ask a faculty member at her university but the supervisor simply look[ed] at the brochures for the two departments and advised, "This one. It looks more beautiful." On this advice alone, Pei-Shan decided which university offer to accept. As the Taiwanese supervisor had no knowledge of the relative merits of the programs in the two universities Pei-Shan had shortlisted, the destination decision was made based on something as extraneous as how photogenic each department's building was. In this manner, interviewees' decisions were not always very "scientific," but combined both strategic and subjective thinking.

The Rise of Asian Scientific Training Programs

While the outsized gravitational pull of the USA as a prime destination for graduate and postdoctoral training is largely expected, this chapter reveals the recent rise of Asian scientific training programs (particularly at the doctoral level and below) that are creating viable alternative training pathways for aspiring Asian scientists. Some of this rise is being driven by the increased marketing outreach that top Asian universities have been undertaking in recent years (following the approach spearheaded by American universities).[35] Two decades into the twenty-first century, top Asian universities are becoming more aggressive in their efforts to lure Asian graduate students to choose Asia rather than the West for their initial scientific training. To compete against well-established and more prestigious doctoral programs in the West, the availability of scholarships is an important lure that Asian universities use to sway prospective international graduate students during their migration and destination decision-making.

The Singapore government has established the Singapore International Graduate Award (or SINGA) scholarship that provides full financial support (without any postgraduation service requirement) to international students pursuing doctoral degrees in the biomedical/physical sciences or engineering in three of Singapore's national universities.[36] In 2002, the Taiwanese government established the Taiwan International Graduate Program (TIGP),[37] an interdisciplinary PhD program through a joint partnership between its premier research institute, Academia Sinica, and several of Taiwan's top research universities, aimed at recruiting the world's best doctoral students in science through competitive scholarships and an all-English learning environment. The TIGP is marketed as cheaper but just as enriching as doctoral programs in Western countries.

I witnessed the impact of these new Asian initiatives through the training pathways of my youngest interviewees. One such interviewee was Benny, a Chinese biomedical engineer who completed his early training in China and was then awarded a full scholarship to pursue a PhD at a Singapore university in a newly launched bioengineering program in the early 2000s. Benny explained to me that he was offered a place in a similar program at a public university in the Midwestern USA, but found the offer from Singapore to be more generous. He also

---

[35] Levin 2010; Cummings 2011; Paul and Long 2016.
[36] "Singapore International Graduate Award," Agency for Science, Technology and Research (A*STAR), accessed on November 27, 2020, www.a-star.edu.sg/Scholarships/for-graduate-studies/singapore-international-graduate-award-singa.
[37] Taiwan International Graduate Program @ Academic Sinica, accessed on November 27, 2020, https://tigp.sinica.edu.tw/.

noted that this Singaporean university's global ranking in engineering was very high, and a degree from Singapore was respected in China, where he planned to return. For these reasons, he turned down the USA offer and completed his PhD in Singapore instead, after which he applied for and received a postdoctoral fellowship at an Ivy League university.

Ambitious Asian universities are actively taking steps to woo Asian students through such scholarship programs as well as through partnerships with top-ranked Western universities. As mentioned in Chapter 2, Singaporean universities have set up partnerships with top universities in the West to offer joint degree programs at the graduate level so that Asian students will be more willing to pursue their graduate education in Asia itself without the fear that they are missing out on an experience at a top Western university.

In another indicator of the rise of Asian science training programs, interviewees shared that, increasingly, the question of where to train had to be decided at the level of the particular university or lab a student was accepted into, rather than on the basis of blanket statements about the superiority of the training in a particular country over another. Unni, an Indian bioscientist who is now a full professor at an Ivy League university, noted that significant progress had been made in India's top research universities. To him, acceptance into a doctoral program at one of the top Indian institutions was markedly different from acceptance into a regional state university in India. As such, in order to answer the question of where to pursue scientific training, one needed to compare the specific universities or labs where one had been accepted, rather than just the country or region:

> At some of these select institutions, for instance, the Tata Institute for Fundamental Research, or the Indian Institute of Science, many of the folks out there were educated over here [in the USA]. They came to the US, they did their PhD, they went back, and they have incorporated much of the same technology, much of the same know-how that we have over here into their laboratories. If a student came to me and said that he or she was able to get into one of these institutions, I'd say that he or she would get a fairly good education over there, and be exposed to all of the stuff that we are exposed to over here. That being said, these select institutions continue to remain quite selective, and you don't have them all over India. So it's only the cream of the

cream that get into these institutions. The rest, they get into fairly mediocre places and, instead of doing that, I would counsel them to come over here and familiarize themselves with the technology that pretty much every reputable academic institution has at its disposal over here.

Unni's advice was a remarkable shift from the advice given in the 1980s and 1990s, where *any* university in the West was deemed better than any university in Asia. His advice speaks to the rise in standing of select Asian universities over the last couple of decades.

There was also growing recognition of the rising status of select Asian labs. Indian biologist Ram, who completed his PhD in India before going to the USA for several years of postdoctoral training, insisted that the comparison between destination options had to be conducted at the level of the lab and the PI, before he could decide what destination advice to give his current students:

> I'll just compare labs. There are crappy places in the US and Europe; there are good labs in India too. So it really boils down to that, especially now that there are many good labs in India. So I would ask him or her which lab in India versus which lab in the US.

Other scientists argued that the decision of where to go for training should be determined by where aspiring scientists imagined their future careers to be based. Yan Jiang, a Chinese scientist working at a research university in the East Coast of the USA, argued that, purely in terms of skills training, doctoral students in China now received better training than students in the USA, but that other career considerations had to be taken into account:

> It really depends on what's their final goal. ... For the career development, I will say they should get their PhD in China, come here [to the USA] for short-term [postdoctoral] training, then go back to China. But if they want to stay here [in the USA], then they definitely have to come here to get their PhD training. Because eventually it's really about your social network and so, if you want to survive in the US, you'd better come here early.

Yan Jiang elaborated on the strength of the scientific training in contemporary China:

> I think the training in China actually, if anything, I think it's better than the training here [in the USA]. Yeah, those people from China, usually they're very skilled. Actually, right now, our lab is looking for a technician or a postdoc. A lot of PIs will prefer people from China because they're well trained. Although their language, their social and presenting skills may not be good, but their technical [skills] are actually much better than people trained from here.

Yan Jiang's advice that young aspiring scientists from Asia should first consider their own career goals and tailor a training pathway that best matched those goals, again speaks to the emergence of multiple viable pathways for Asian scientists. Similar advice was shared by returned Taiwanese scientist, Li Wei. She explicitly tells her present-day students in Taiwan that they now have multiple training options to choose from, rather than the straightforward "Go West" directive that earlier generations of Asian scientists (like Li Wei herself) used to receive:

> I always tell students, "You have different ways to pursue your career. You can finish your PhD very quickly [here in Taiwan], because you don't need to adjust to the culture, you don't need to polish your language. So you can finish your degree within five to six years and then you go do postdoc in the US or Europe. Because you need to have an international exposure. And that will be great. But you can also go to US or go to Europe for your PhD, but it will take a much longer time. But once you finish the degree, you have already polished your language, you know what's going on in outside world, so it's fine too. So I don't have an exact advice but I can explain this to them.

Li Wei's parting words that she does not have "exact advice" highlights the shifts that have occurred in the Asian scientist migration system in the last couple of decades, which speak to the shifting topography of the global scientific field.

Conclusion

This chapter explored the self-reported motivations of the 119 Asian scientists I interviewed who moved to various countries in the West for training, starting in the 1960s all the way into the 2000s. This

Conclusion 113

chapter depicts how these Asian scientists understood their migration and destination options, how they went about their decision-making processes, and who (and what) was influencing their decisions at that time in their lives. Their migrations to the West occurred within the broader context of their developing sense of themselves, their ambitions, and their evolving understanding of the best places in the world to pursue those ambitions. They also made their decisions in an environment of imperfect and incomplete information, and these decisions were as much subjective as they were strategic.

In the second half of the twentieth century, interviewees' desire to travel to the West was over-determined. There was no question that graduate training in the West was superior to what was available in Asia. By the 2000s, however, the Asian scientific training environment had improved significantly, particularly in the top universities in my four Asian case countries. There were now more reasons to pursue initial training in Asia and delay migration to the West until later in the career course. Exposure to Western approaches to scientific research was still considered important but this preference was increasingly using a lateral understanding of the benefits of brain circulation rather than an assumption of technological inadequacy on the part of top Asian research institutions. At the postdoctoral training moment however, there was more agreement that this phase of an aspiring scientist's training should be conducted in a Western country and preferably the USA. There was an acknowledgement that the core networks in the life sciences were still concentrated in the USA and there was no real substitute (as yet) for being physically embedded in those networks for a few years.

By 2010, a large segment of my interviewees had returned to Asia where they were pursuing independent scientific careers in top-tier research universities or institutes. When giving training advice to their own students, these returning scientists were now more informed and more strategic in their suggestions. They raised the issue of the science-related social and cultural capitals associated with Western training that could be converted into symbolic capital upon their students' return home. Several argued that the governmental bureaucracy and scientific community in their home countries continued to privilege scientists with Western qualifications over all others. Most interviewees were convinced that the recognition and reward structures of the academic scientific field within their Asian home countries required aspiring scientists to acquire some overseas (and primarily Western) qualifications if they wanted to rise within these fields.

These arguments demonstrate how the reasons for seeking scientific training in the West have shifted since the twentieth century. The motivation to train in the West was now less about Western universities' relative technical expertise and more about the scientific cultural and social capital that the West bestowed upon an aspiring Asian scientist who trained there. But, as I discuss later in this book, returned scientists also felt the need to keep their best and brightest students in their own labs in Asia, so that they could benefit from these students' abilities and energies for a longer period. As more and more ambitious Asian scientists return to work in Asia, this pressure to keep trainees in Asia will only increase, cementing these changes in the Asian scientist migration system by altering the calculus regarding the appropriate time to travel to the West.

4 | *Learning Science in the West*

Interviewees who arrived in the West between the 1960s and the 1990s typically found themselves in a scientific research system of material plenty, compared with what they had experienced earlier in their Asian universities. Interviewees from China and India in particular remarked on how they joined Western institutions that were awash in advanced research equipment, generous research funds, and supported by efficient administrative processes. Manas, who had previously trained at the Indian Institute of Science (IISc), which he considered "one of the best research institutes in the country," said he was "blown away" by the research infrastructure he encountered in the USA when he first arrived in the 1990s.

Interviewees worked hard to make the most of these new research opportunities. Eng Chye, a Chinese scientist who moved to Western Europe, argued that because he had lived through the Cultural Revolution as a child, he did not want to waste any of the possibilities he encountered overseas. "We have gone through hard times, so we know once we catch a good opportunity," he explained.

While the stark material differences between my interviewees' Asian and Western training institutions were important, this chapter focuses instead on the differing cultural values and norms my interviewees were exposed to regarding the social practice of scientific research. Arriving at top universities and research institutes in the West when they were still relatively junior in their scientific careers (see Table 4.1),[1] interviewees were exposed to not only new domains of scientific knowledge and new scientific techniques and technologies but also new scientific cultures. During their interviews with me, they talked at length about how this exposure reshaped their normative ideas about the best way to approach the teaching of science and the optimal composition and management of

---

[1] Note also Table 1.1, which lists the most common research universities/institutes where interviewees trained in the West.

Table 4.1 *Number of interviewees by Western destination, first training stage in the West, and institution type*

Country/Region of training	Institution type	First training stage in the West			
		Undergraduate	Master's	PhD	Postdoc.
USA	Ivy League universities	–	3	10	4
	Public universities	3	6	36	7
	Private universities	3	1	13	6
	Research institutes	–	–	2	4
Canada	Public universities	–	1	2	–
UK	Public universities	1	1	1	–
	Oxford/Cambridge*	–	–	4	2
Europe	Public universities	–	2	5	1
	Research institutes	–	–	1	–
Total		7	14	74	24

Note: N = 119. The number of postdoctoral training institutions reflect the fact that some interviewees held multiple postdoctoral fellowships, often in different institutions.

* Oxford and Cambridge are technically public universities in that they receive government funding and operate under state regulation about the fees they are allowed to charge students, and various reporting requirements.

scientific teams. It is this new set of values, priorities and approaches to scientific teaching and research that I dwell on in this chapter.

One reason why these scientists may have focused less on the technical and more on the cultural aspects of their new scientific environments in the West was because they were speaking with someone who was not trained in the life sciences. In this regard, I benefited from my "naïve outsider" status. Interviewees did not tell me about the technical differences between the instruments, research supplies and model systems they had access to in the West vis-à-vis Asia. They did not talk at length about their research methods or their particular research subspecializations, though we did touch on these topics during the interviews. Instead, they initiated conversations about differences in how science subjects were taught, and how labs were managed across different places and organizations. They described the power of scientific cultures to shape the environments in which research goals are established, research teams constituted, and research norms and values inculcated. It was this particular emphasis in our conversations that led me to focus this chapter on the differences in scientific cultures across labs, universities, institutes and even countries.

Scientific cultures are complex and contingent sets of norms and values, shared within a given scientific community, about the social practice of scientific teaching and research. While the scientific method refers to the iterative and systematic approach scientists adopt when asking specific research questions, theorizing possible answers, and drawing conclusions from observations, the scientific culture of a science class or lab team speaks to the broader normative environment in which scientific subjects are taught and the scientific method is pursued to answer particular research questions. It is akin to an organizational culture that "constrains all other aspects of organizational life and limits what is considered desirable, possible and practical to do."[2]

Using a grounded, inductive process, I identified seven dimensions of scientific cultures from my interviews.[3] This chapter details each

---

[2] Smith and McKeen 2011:4; Kotter 1996.
[3] Not all interviewees spoke of *all* of these facets of scientific research cultures during their conversations with me. I constructed my taxonomy by collating and organizing the terms and frames that different interviewees organically used, rather than relying on any preconceived set of dimensions I systematically asked each interviewee. In future work, I hope to conduct large-scale surveys of

of these dimensions using interviewees' comparative accounts of their early training in Asia and subsequent training in the West. Interviewees provided me with rich, evocative descriptions of the differences between the scientific cultures of their home-country institutions and those in their Western training institutions. Studying and working in Western countries had a significant impact on how my interviewees envisioned themselves as practicing scientists, how they wanted to work with and supervise others in their own lab, and the relative importance they gave to questions of diversity, communication, critical thinking, and egalitarianism in research settings. This impact is worth understanding because it influenced interviewees' return decision-making, as I discuss in the next chapter. For those who chose to return, it also affected how they aspired to teach science, conduct scientific research, and mentor the next generation of scientists in Asia.

This chapter also documents the significant variation in each of these dimensions not only *between* Asia and the West, but also *within* each of these world regions at the level of countries, universities, institutes and also individual labs. The essentializing language of "Western science" and "Asian science" was used by many of my interviewees without any prompting from me, but I try to avoid such language as far as possible because I do not want to create the impression that a monolithic scientific culture exists in the West or in Asia.[4] In fact, even as they used terms like "Western science" or "Asian science," my interviewees routinely highlighted differences between different parts of the West (particularly between the USA and Europe), between universities within a single country, and even between labs run by different professors in a single university. By deconstructing scientific cultures into their seven constituent dimensions, I also show that not all of these dimensions change in the same direction as one moves across regions, countries or institutions. Even as I describe high-level differences in scientific cultures between Asia and the West, I document how

---

students and trainees in various research institutions across Asia to develop a more comprehensive picture of the variations that exist in scientific cultures in different Asian locations and the factors that influence this variation.

[4] Scholars like Richard E. Nisbett (2003) have made such essentializing statements, arguing that "Asians and Westerners think differently." They have been roundly criticized by anthropologists like Sherry Ortner (2003) who highlight the problematics of thinking dichotomously about "Westerners" and "Asians" in this manner.

Defining Scientific Cultures 119

different countries within either of these world regions are varyingly situated on each dimension.⁵

Finally, this chapter highlights how scientific cultural norms are learned and diffused. Interviewees did not simply stop at observing differences in scientific cultures across their Asian and Western universities. They made value judgments about which aspects of these cultures they personally preferred, and then adopted and adapted their own norms and preferences accordingly. This last takeaway is perhaps the most important one of this chapter because it demonstrates how there is nothing innate or fixed in the scientific cultures of various Asian countries. The Asian scientists I interviewed changed their approach to scientific research and teaching as a result of their time training in the West, and those who returned to Asia carried these changed values and norms back with them and tried to inculcate them in their new classrooms and new labs in Asia. This was a key part of the "scientific remittances" these returning scientists brought back with them, as I discuss in Chapter 8.

Defining Scientific Cultures

When Robert Merton identified the four norms associated with science – universalism, communality, disinterestedness, and organized skepticism – which he deemed to be universal principles that scientists internalize as their "scientific conscience," he was speaking as if there is a common value system shared by all scientists.⁶ However, later scholars have demonstrated that there are in fact *many* scientific cultures and that they can be highly variable.

There exists much scholarship within STS on this topic of scientific cultures, typically based on studies of a single lab or research culture.⁷ Pierre Bourdieu's notion of a scientific habitus parallels that of a scientific culture internalized by an individual scientist. To

---

⁵ I do note however that most of my examples about scientific cultures in the West are about top-ranked American universities, which are not representative of the West as a whole or even of the United States. While I tried to include as many comparisons with European universities as possible, I acknowledge a shortage of data, which largely stems from the fact that most of my interviewees trained in the USA. I hope that my future research on scientific cultures around the world will correct this imbalance.
⁶ Merton 1973; Barber 1962; Storer 1966.
⁷ Franklin 1995; Traweek 1993; Krautwurst 2014.

Bourdieu, the scientific habitus encompassed the "set of attitudes which structure the perception and practice of research" and "[relate] to the way in which the problems are set out, explanations developed and tools forged and used."[8] Meanwhile, Londa Schiebinger writes that scientific cultures encompass "rituals of day-to-day conformity, codes governing language, styles of interactions, modes of dress, hierarchies of values and practices" (1999:68). In describing the scientific culture in the neuroendocrinology lab they were studying, Bruno Latour and Steve Woolgar refer to it as "the set of arguments and beliefs to which there is a constant appeal in daily life and which is the focus of all passions, fears and respect" (1979:55). Like Latour and Woolgar, many STS scholars have studied cultures of science by conducting in-depth ethnographies of research laboratories, detailing how the research that is carried on inside these scientific spaces is influenced by a range of factors (from sociodemographic variables, to particular value-systems, to specific models of the natural world) and not just scientific "facts." Latour and Woolgar described the culture of their neuroendocrinology lab as held together by the belief in a particular model of brain function. Given the wide range of subfields within the life sciences that I am studying, I cannot talk about the valorization of particular techniques, theoretical models, or model organisms, as these vary from subfield to subfield. Instead, I focus on the values and norms surrounding scientific teaching and research management and organization, to look for commonalities as well as patterns of variance across different locations.

This requires me to first map out the constituent dimensions of any given scientific culture. Focusing primarily on the "hierarchies of values and practices" (Schiebinger 1999:68) within scientific communities, I identify seven "value and practice" dimensions from my interviewees' accounts of the scientific cultures they encountered in each of the institutions and countries where they trained. Typically, the scientific cultures of particular countries are written about in essentializing language that highlights only one dimension as the most important factor in understanding how science is practiced in that country. For example, the scientific culture in Chinese research universities has been described as "animated less by science entrepreneurialism than by lab

---

[8] Lenoir 2006:27.

Defining Scientific Cultures

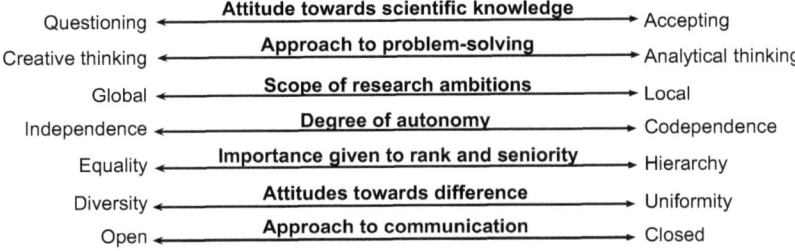

Figure 4.1 Seven dimensions of scientific cultures

drudgery" (Ong 2016:126). Meanwhile, Singaporean institutional research culture is characterized as motivated by "Singaporeans' deep-rooted aversion to failure" (Wong 2011:178). I aim to go beyond these somewhat one-sided observations by revealing the multidimensional nature of all scientific cultures.

Each dimension I identify references a particular aspect of scientific teaching and research practice. Each of these dimensions is imagined as existing on a continuum with two opposing poles. I frame each of these dimensions in language that is as neutral as possible so that there is no implicit judgment on my part as to which end of each spectrum is preferable. However, as the remainder of this chapter will demonstrate, interviewees *did* express clear preferences of their own about the kind of scientific culture they wanted to instill in their own labs. The seven dimensions of a scientific culture that I identified were (see Figure 4.1):

(1) The *attitude towards scientific knowledge*, which ranges from having a critical attitude towards new and existing knowledge, to encouraging the acceptance of received wisdom.

(2) The *approach to problem-solving*, with one end prioritizing out-of-the-box, creative and disciplinary-spanning idea generation, and the other end prioritizing more analytical approaches that logically break down a problem into its constituent parts.

(3) The *scope of research ambitions*, which can vary from an aspiration to be the best in the world in a particular domain to more locally focused ambitions.

(4) The *degree of autonomy* given to individual scientists, with some environments requiring independence and others emphasizing codependence.

(5) The *importance given to rank and seniority*, with some cultures emphasizing equality between individual scientists regardless of rank, while other environments prioritize respect for hierarchy and seniority.
(6) The *attitude towards difference*, with some environments geared towards accepting individuals from highly diverse backgrounds whether along nationality, ethnicity, gender or disciplinary lines, while other environments value uniformity within groups.
(7) The *approach to communication*, with some scientific cultures prioritizing a broad and open information-sharing approach that emphasizes the importance of outward communication to non-members as well as inward communication among team members, while other cultures take a more closed approach that emphasizes sharing data only on a need-to-know basis.

For each of the above dimensions, different countries, institutions and labs may be placed at different points on the relevant spectrum. I try not to make sweeping generalizations about a monolithic "Western scientific culture" or an "Asian scientific culture." It is true that interviewees tended to place the scientific cultures in the Western institutions where they trained towards the left end of the spectrum for most of these seven dimensions, as compared to the scientific cultures of their Asian educational institutions. However, within the West, interviewees positioned top-ranked private American universities more to the left as compared to public European institutions, particularly when it came to the scope of their institutional research ambitions and their attitudes towards difference. Within Asia, Singaporean universities were held up by several interviewees as manifesting a greater tolerance of difference, compared to institutions in other Asian countries, while all four of my main Asian country sites – Singapore, Taiwan, China and India – were highlighted as places where hierarchy mattered greatly. Several Indian scientists spoke of the extreme variation on all seven dimensions that existed *within* their country, favorably comparing top national research institutes, such as the National Centre for Biological Sciences (NCBS) in Bangalore, against regional/state universities in India. For these reasons, I use the plural term "scientific cultures" throughout this chapter to highlight how *many* cultures can exist within a given world region or country.

I also caution readers that each of the seven dimensions I describe here are highly complex, multifaceted and could be easily further subdivided into more narrow subdimensions. A scientific culture's approach to communication, for instance, involves both a group's internal and external communication approach. It can also reference its norms around oral as well as written communication. In fact, it is possible to talk about a group or organization's knowledge-sharing norms as an information "culture" in and of itself. However, to keep things simple, I only highlight some of these nuances in my discussion of each dimension, leaving it to future research to go into further detail about each of these dimensions.

Learning New Scientific Cultures

Attitude towards Scientific Knowledge

John Dewey, the influential American education reformer and psychologist from the first half of the twentieth century, wrote at length about learning as a social process, with learners influenced by the broader social and cultural environment in which they operate. To Dewey, asking questions was innately human, but this trait could only be nurtured by an inquiry-driven learning environment rather than one that emphasized rote learning.[9]

Interviewees regularly spoke about how the approach to learning scientific knowledge that they encountered in their Western universities was much more inquiry-driven and grounded in critical reflection than what they had been exposed to in their home country. Richard, a Singaporean clinician scientist, spoke of how revelatory this had been for him when he was first exposed to clinical research in the UK. This was an approach to the practice of medicine that he had never encountered when he trained to be a doctor in Singapore in the early 1990s:

> At the time of being in medical school [in Singapore], it was very clinical training, you know? They're training people to be practitioners. [But the UK] gives clinicians a window to see that there are ways to explore the frontier. Medicine is very much like cookery. We follow a cookbook, guidelines laid down by others,

[9] Dewey 2004 [1916].

and then you dispense the medicine. Whereas in research, it's frontier work. You are actually trying to push and find out new things. And I really enjoyed that.

Interviewees argued that this kind of intellectual curiosity was deeply embedded within the cultures of the various Western educational institutions where they trained. A handful of interviewees had already experienced this kind of curiosity-driven learning environment in their home country in Asia – in a particular department or in an encounter with a particular professor – but they found that it was much more pervasive in the West, where they experienced it as an institutional environment. As a result, several spoke of undergoing a scientific culture shock when they first started studying in the West. Taiwanese bioscientist Yi Fang shared that her educational experiences in Taiwan had not prepared her for the critical thinking she was expected to demonstrate in her US university, where she underwent graduate training:

> The education here [in Taiwan] is: you follow what the teacher says. It's quite restricted. But there [in the USA], it's quite creative thinking and everybody can criticize the paper. The first time we have that discussion class, I really cannot criticize the paper! Because [in Taiwan,] we read a paper like the Bible. And that difference – you struggle with it a little bit. But I started to pick it up. I think that's quite [important] if you want to stay in science.

Likewise, Liu, a bioengineer from China, described the USA as having an "established system that encourages research, encourages people's curiosity." Liu shared how "crazy" he found the "freedom" to ask questions when he started his PhD in the USA:

> I expected some difference because I knew that, in Boston, the people are doing really top research. But I didn't expect you're free to do any research over there! I mean, just no restrictions. That's what I didn't expect. You're free to talk to anybody and there's all kinds of crazy people there. That's what I didn't expect. I mean, people like to talk about research always! They're ready to do anything!

Liu made a key distinction between pursuing "top research," which he understood as asking high-impact research questions, and being "free" to pursue any research question that interested him, no matter how small or arcane the question might be. He had expected and desired the former, but was surprised when he encountered the latter. He subsequently came to believe that there was a causal relationship between the two, with research freedom driving research impact. Meanwhile, Chris, a Singaporean neuroscientist, contrasted the American educational system's open inquiry approach with what he considered the narrowness of the education system in Singapore, which he had experienced as more focused on the immediate applications of new knowledge:

> A lot of [American] culture upholds, and is aware of, the value of research. Not just in terms of what it can do for us, but just for the pure joy of knowing and finding things out. That's really what they have there that I can really appreciate and that really struck deeply in me.

Interviewees regularly ascribed the lack of an inquiry-based culture in their home country's educational institutions to the cultural norm of not questioning one's elders. Manas, who went to the USA from India for his doctoral studies, remarked on how students in India "never asked questions," as opposed to students from the USA. From his perspective, this was because the learning environment in most Indian educational institutions did not facilitate or reward a critical stance towards existing knowledge. Similarly, Hong Joo, one of my Korean interviewees who trained in the USA, described the process of writing his undergraduate thesis in Korea as: "You go to the lab, do some kind of washing dishes and cleaning up benches, and then you get some data from the master's students, and kind of copy-and-paste and just submit." Yong Kai, a Taiwanese molecular biologist, described how young people in Taiwan "are told just to listen, don't try to speak from your mind. You know, just follow the order. But in the USA, they always ask you, 'Why?'"

Scientists from all four Asian case countries bemoaned how their home-country scientific training had lacked a curiosity-driven approach. Nishith, who completed his undergraduate, master's and doctoral degrees in India in the 1980s and 1990s before moving to the USA, described his undergraduate studies in physics as having been completely based on memorization of the material in his textbooks and all his assessments being via written examinations. There was no research component to his studies at all:

> So there is no research. Of course, you may have to give some research talk once in a while, but again that is college-dependent. Not in every college would they try to promote or encourage such things. So in my college, my professors used to tell each student to give a talk. But I wouldn't go too deep into any research. ... There was no journal or anything like that in my small college. So with the available resources of textbooks, you try to prepare something to give a talk for ten minutes.

But Nishith stressed that this talk was simply summarizing what was in the textbooks, rather than offering a critical analysis of the material or posing possible new questions to explore. Likewise, Ram, an Indian scientist who went to the USA for postdoctoral training, shared that while he received a strong technical foundation in India, he was not taught what he termed "the *whys* of anything":

> I went there [to the USA] and started realizing that getting papers published is different than thinking about them and coming up with a good research problem. And that was an eye-opener because you would start talking to people and then they would ask questions and you were like, "Oh, yeah! I never thought about that." So it took some serious time and often it was very embarrassing and, you know, it didn't come naturally.

While challenging, Ram was convinced that this rewiring of his brain to learn how to ask more critical questions had had a positive impact on his subsequent scientific research.

Linked to this culture of questioning in the USA was a culture of seeking *answers* to these questions, rather than simply accepting that some questions were unanswerable. Yiu Man from China praised the early introduction to hypothesis construction and testing that occurred in Western schools. He joked that, now that he was back in China, he would be more useful to his birth country if he taught in primary schools because he could affect more fundamental change at that early age:

> [In the USA,] they ask questions from a very young age, which we do not do here [in China]. We are asked to read exam materials and try to perform well, according to the questions

shown on the paper, so it's very different. So our education system does not ask us to ask questions, to observe and to form hypotheses and to devise a method to address the hypotheses, and to find out – to have the desire to find out truths. That's the culture of things [in the USA].

This dream of changing the culture of learning back in the home country was a key motivator behind some interviewees' desire to return home. I speak more about returnees' experiences trying to effect such changes in Chapter 8.

Approach to Problem-Solving

Another recurring theme raised in interviewees' accounts of their experiences in Western educational institutions was that the environments in these institutions tended to encourage more creative problem-solving, including a broader acceptance of failure as simply part of the innovation process. Mathematician Marcus du Sautoy (2019) cites artificial intelligence theorist Margaret Boden in describing three types of creativity: exploratory creativity (which pushes the limits of defined fields), combinatory creativity (which fuses or plays with the parameters of two different domains to produce something new), and finally, transformational creativity (which is the hardest of all because it involves completely changing a field).[10] Du Sautoy notes that deep and rigorous training in one's field is absolutely essential for creativity to flourish.[11] But he also advises that an environment that encourages exploration and training *outside* one's field is helpful in fostering combinatory creativity. Likewise, an environment that permits the questioning of rules and assumptions of a given field of study (after first gaining familiarity with these rules) can encourage transformational creativity. It was in all these areas that interviewees felt that their training institutions in the West, and particularly in the USA, were strongest.

Several interviewees hypothesized that the cultures of creative problem-solving they encountered in the West were linked to the breadth of education that Western students (and especially those in the USA) received. Interviewees shared that graduate students in the

---

[10] Boden 2004[1990], 2016.
[11] Du Sautoy 2019:15.

USA were introduced to a range of ideas from a variety of disciplines, allowing for easier cross-fertilization and, therefore, the potential for more combinatory creativity. In contrast, Chinese biochemist Larry spoke of how much more focused, and therefore more narrow, he found his training in China. He described this as a detriment to scientific creativity in the long run:

> If I was in a chemistry department [in the USA], I will be encouraged by my professor to read many different books in chemistry. But, in China, for PhD students, usually you are supposed only to read the books directly related to your project. ... So the consequence in China is, if you're sitting in a seminar, usually few students want to ask questions, because if the speaker's research area is slightly different from the students' research area, the students don't know what to ask. Because they know very little about the new research field. But in the USA, usually the speaker in a seminar will receive many different questions from students, because students are supposed to read more about the *whole* area.

Hock Peng, a Hong Kong-born scientist now working in Singapore, described the US-based research teams he had been part of as more willing to consider out-of-the-box ideas: "The group I joined – we would just like throw ideas at each other. We talk about it whenever we have an idea or just want to talk about the science. 'Hey, how about this? Oh, that sounds like a cool idea!'" In contrast to the exploratory creativity that marked his US experience, Hock Peng felt a "barrier" that made him feel "slightly less creative" in Singapore, describing a conference he had attended after moving to Singapore where he felt that there was "nothing very spectacular or there's no new novel idea that excites me. ... It wasn't very interesting." Likewise, Sai Kiang noted that the science education he had received in China was overly focused on error reduction rather than creative exploration:

> My experience with research in Beijing was ... repetition of the same thing over and over again. ... Hypothesis-driven research is never here in China. ... I think professors here at the time were just paying more attention to not make mistakes and be careful. That kind of approach does not encourage you to think big picture.

Learning New Scientific Cultures

Openness to broad-based learning and the free exchange of ideas across disciplines and projects was not limited to the USA however. Runchen, a Taiwanese biochemist who trained in the UK, spoke of how there was a tradition of daily informal gatherings at her department:

> Teatime is important! [*Runchen laughs*] In [my UK university], everybody would come properly in the morning, 10:30, because that's early morning time – the morning tea. And 3:30 to 4:00, that's afternoon tea or coffee break. And that time is a good time for exchange of ideas. I think that's very important. ... You'll just chat and share, not a formal presentation ... You'll find, in the UK, tea break everywhere!

To Runchen, this relaxed culture of informal sharing – which is closely linked to the communication culture of a research organization – allowed for new ideas to be generated. Rather than guarding their ideas, researchers chatted with members of other labs in the relaxed setting of the tearoom about their respective projects, leading to the generation of new research possibilities and collaborations.

Some interviewees did encounter such cultures of curiosity in their training institutions in Asia, but this was rather rare within my sample. Divya, an Indian ecologist, described to me a summer science camp at one of India's top science universities that she attended as an undergraduate. She shared how that summer experience motivated her to pursue a career in science because it exposed her to a discovery-driven approach to scientific research:

> In the atmosphere [at the summer camp], you felt free to do what you wanted. If you went up to this professor and ask, "So why do you think bacteria are colorful?" he'd say, "You tell me."
>
> And so you came up with your favorite hypothesis, and then you ask him, "Do you think that's true?" Because that's what we're used to thinking: that professors know everything, and they know the answers, and they will give us the answers.
>
> But he's like, "I don't know. You should find out. Why don't you find out?"
>
> And so then you went and did some experiments in the lab. So you were perfectly free to go and potter around, and design your experiments, and do your experiments without having approval from anyone. There was no signing out of anything. There was no bureaucracy. No red tape. So you just go and you say, okay, I'm going to clean up these many glass petri dishes that I need for

this experiment. I will autoclave them, sterilize them, get all the media ready whenever I want, and then I do my experiment.

I didn't realize how unique that was until [I] got out. Then you say, "Oh, wait, no one else got that [in India]!"

Divya felt the difference between the problem-solving cultures of her summer science camp and her undergraduate university, though both were in India. It was the summer science camp that introduced her to a different way of approaching scientific questions. That early exposure in India ensured that when she went to the USA for graduate training, she was able to easily adjust to the similarly creative, open-ended process of scientific experimentation and discovery that pervaded her doctoral classes and labs.

Scope of Research Ambitions

The third area where interviewees experienced a difference in scientific cultures was in the scope of professional ambitions encouraged and nurtured by their research organization. Interviewees recalled what it had felt like to arrive at elite universities in the West and rub shoulders with the authors of foundational textbooks they had read as undergraduates or papers that were considered classics in their particular subfield. They met scientists who demonstrated a level of ambition they had rarely encountered during their early years in Asia. Arjun Appadurai has written that "aspirations are never simply individual. They are always formed in interaction and in the thick of life" (2004:67). In moving to a Western country, interviewees changed their social milieu and encountered new peers and professors who harbored great research ambitions for themselves, thereby expanding interviewees' personal "aspirations window."[12]

As a result of this cultural shift, some interviewees initially questioned their ability (as well as their desire) to operate at this higher level. For interviewees who had regularly been at the top of their class back in their home country, this moment was sometimes the first time they felt truly stretched on an intellectual level. Weiliang spoke about his

---

[12] Economist Debraj Ray defines the aspirations window as "formed from an individual's cognitive world, her zone of 'similar', 'attainable' individuals. Our individual draws her aspirations from the lives, achievements, or ideals of those who exist in her aspirations window" (2006:410).

Learning New Scientific Cultures 131

experience moving to a West Coast university considered one of the best in the world, and finding himself hitting a wall for the first time in his life:

> I know I have the ability to learn but, in [my US university], I take many classes and I feel the boundaries of the things that I could understand. [*Weiliang pauses*] And that's sort of a check for me. Actually, in my whole life, I never felt that way before. I always come to a setting where I can do it, but in those [US] settings, I did feel that there is a limit to my potential. Many of [my teachers] are so knowledgeable and read so widely, then you feel that even if you spend your whole life reading you cannot catch up with them. But I really liked the exposure to the whole setting. I think that really opened up my mind.

Despite this challenging adaptation he underwent during his initial months in the USA, this new environment stoked Weiliang's ambitions and encouraged him to aim higher.

It was not only the faculty they were training under who inspired soul-searching and expanded ambitions among the scientists I interviewed. Interviewees' fellow doctoral and postdoctoral trainees often had a similar effect. Encountering other brilliant students, Weiliang recognized that he needed to work harder if he wanted to keep up. As he put it, "I really started to feel that I have to be serious. Basically the [US] experience was a reality check for me. And I also really understand that I have to be responsible for myself."

As a result of being surrounded by more ambitious peers and supervisors, interviewees often began to expand their own research ambitions as well. Dodi, an Indian scientist who trained in the USA, contrasted the American and Indian ways of thinking as "thinking really big, rather than thinking in small steps." Similarly, Shirlena recounted how her scientific ambitions expanded during her postdoctoral training in the USA, even though she had thought they were already quite grand when she was still in Taiwan:

> [In Taiwan] the student maybe will get some more publications compared to his friends, so he thinks he's very good and he has very high confidence. And he thinks he knows everything. But this view is so narrow. If they don't go to another country, they will think they are the king, they are the best. But it's not real. And also the research direction will be only limited to what he

> already knows about. Compared to me, when I was [in Taiwan], I got five papers published, so I was very good. But... when I go to US, I find that the research is totally different.... So my vision has become bigger.

In Shirlena's case, her "vision" of what she wanted to achieve with her career became so big that she decided she needed to return to Taiwan after her US training, where she could make a name for herself away from the shadow of her American postdoctoral supervisor:

> My lab supervisor [in the USA] was the king. He was the king. He is the biggest.... We were living under his shadow. Nobody knows about us; they only know about [him]. And I was scared that I couldn't establish my own lab, do my own thing in the US. But I also want to switch; I want to do something else.... Even though I don't say out loud my dream, but I think it's very important in my heart that I have to achieve the dream.

Shirlena returned over the objections of her American supervisor, who warned her that she was committing career suicide by going back to Taiwan. But Shirlena suspected that her supervisor simply wanted her to continue working in his lab because of how productive and capable she was. In Shirlena's case, her expanded ambitions led her to decide to return to Taiwan, where she knew she would be given a higher rank, larger research funds, and more status by her Taiwanese institution than what she would receive in the USA with her former supervisor taking up most of the spotlight. But earlier cohorts of Asian scientists tended to choose to stay in the West because they saw greater scope to realize their research ambitions in the West rather than back home. However as the scientific research systems in Asia began to improve, the latest cohort of my interviewees was much more willing to consider return. In the next chapter, I dig further into how their expanded ambitions actually encouraged more return decisions.

Interviewees were also aware that not all research organizations in the West operated at the same level when it came to the scope of their institutional ambitions. Most interviewees were of the opinion that scientists at American universities were more ambitious (or "pushy," in the words of one interviewee) than European ones. This scientist, Jialiang, who first left China to train in a European country, highlighted the difference in scientific cultures between this country (where he completed his PhD), the USA (where he completed a postdoctoral fellowship), and China (where he completed his undergraduate training and to which he returned after more than a decade overseas):

> In [the European country], people are more lazy. Because it's Europe, they want more freedom and relaxation. In the summer, you can find that people just left [the lab] at four, for no reason – just they want to go to the beach and swim. That's it. And the bosses are also not that pushy. It's very relaxed and not pushy. I think the US is much more pushy and, in China, [now] it is also very pushy ... That's why most of the time, [only] I'm there [in the lab in my European university]. It's very rare to find a [European] student come there on the weekend.

Jialiang's observation that the research culture in China was now as "pushy" as it is in the USA is another indication of how scientific cultures do not fall neatly on an East–West continuum, and how they can change over time.

Several interviewees also distinguished between regional public universities and top private universities in the USA. Soon Huat, from China, spoke about the marked difference he experienced moving from a state university in the American Midwest (where he completed his PhD) to an Ivy League university for his postdoctoral training: "In [the Ivy League university, the] people – students and lab mates – are just brighter. That university is just harder and they are more confident. I think that is the major difference. And the funding resources there are much better than [at the state university]." Soon Huat enjoyed the challenge of operating in the higher stakes environment of his postdoctoral training institution, even as it took him time to adjust to its faster pace and greater competitiveness.

Working and studying in highly productive teams in the West, interviewees also learned the importance of recruiting capable doctoral students and postdoctoral trainees in order to maintain their desired level of research productivity. Because interviewees like Shirlena had contributed to the research productivity of their supervisors in the West, they understood how essential it was to have driven and capable people working for them when they became independent academics and set up their own labs. As I discuss in Chapters 7 and 8, recruiting talented team members is an ongoing struggle for returning scientists in

Asia, as is motivating these research staff to aim as high as possible.[13] In fact, many interviewees from my two earlier cohorts – those who left in the 1990s or earlier – decided not to return because they worried they would lose the ability to regularly interact with talented and driven students, peers and superiors, which they saw as essential for keeping themselves ambitious and motivated.

However, this view of Asia as less driven was already being complicated among my third cohort of interviewees who left Asia in the 2000s. As Chapter 2 showed, my four Asian case countries were pumping large amounts of funding into STEM fields so as to boost their international research profile. Several interviewees shared that they had returned to Asia because they believed their scientific ambitions were better served in top Asian research universities, as the funding there was more generous than what was readily available to them in the West. In this manner, there is now significant variation in the scope of research ambitions one can observe across the contemporary scientific landscape in Asia. Interviewees seemed to self-sort over time, gravitating towards the country and institution that best matches their own personal level of ambition at a particular stage in their career. But this sorting was no longer neatly falling into an East–West divide, and was instead much more attuned to individual institutions' funding packages and different countries' research systems – as I show in Chapter 5.

Degree of Autonomy

Interviewees also argued that there were significant differences across countries and world regions in terms of the degree of individual autonomy expected from and given to trainee scientists, whether doctoral students or postdoctoral trainees. According to interviewees, the scientific cultures in Western countries tended to encourage independence of thought and action among scientists-in-training from a very early stage. They linked this approach to what they saw as the more individualistic cultures in Western countries. But they also argued that this independent thinking was essential for innovative scientific research to flourish. Sarita, an Indian microbiologist, shared her opinion that "individualistic culture fits the culture of science much better. Because you have to be independent, you have to make up your own mind, you have to take

---

[13] See also Ong 2016:124–129.

your own risks." There is in fact a significant body of scholarship that supports this claim, finding that countries with higher rates of innovation and creativity also tend to score higher on individualism measures.[14] The underlying argument here is that "autonomy, independence, and freedom – beliefs associated with individualism – are needed for a nation to be creative" (Rinne et al. 2013:129).

Upon arriving in the West for their training, interviewees recalled learning to fend for themselves in ways they had never had to in their home countries.[15] This was true in terms of needing to adjust to the broader societies they entered, but also to the specific labs where they worked and the classes in which they enrolled. Interviewees were encouraged by their supervisors to generate their own research projects and also take the initiative to troubleshoot any problems they encountered. When he started his first lab rotation in his American doctoral university, Weiliang remembered being told by the professor who would later become his doctoral supervisor: "You have your problems; I have mine." Weiliang took this statement to mean that his supervisor was not going to solve Weiliang's research difficulties for him. Instead, Weiliang would have to find his own solutions to any roadblocks he encountered over the course of his research. Taiwanese scientist Yong Kai was told by his supervisor, "I'm not your babysitter" – which he understood as indicating that his supervisor was not going to provide constant oversight of his research progress.

Interviewees found this cultural shift challenging but also invigorating. Chinese scientist Liu spoke of how his professors in China had viewed their doctoral students as "tools," rather than eventual peers:

> The professor has an idea and he wants you to just do the practical things to realize this idea. This is more likely the style in China. In the US, it is more likely a professor will ask you every day, "Do we have any new ideas?" At the beginning [in the USA], I was kind of confused. Like, you expect everything from me? I thought the professor should design a project *for* me, but actually he didn't. And I think that's very good.

Likewise, Sarita recalled how her American doctoral advisor encouraged her to be independent from the beginning: "She didn't even give

---

[14] Shane 1993; Rinne et al. 2013, 2012.
[15] Migration has often been associated with the inculcation of independence in the individual migrant.

me a PhD problem. I had to find one myself within the context of the lab and get things done. It was fantastic! It was just absolutely fantastic. It made me very independent." Like Liu, Sarita appreciated how the American doctoral training system encouraged students to pursue their own research projects rather than follow or support their supervisor's preestablished research question.

In several cases, interviewees intentionally went to the USA to *seek out* greater independence. Hsing-Yi, a Taiwanese scientist who moved to the USA for doctoral training, disaggregated the impact of her migration from the impact of studying in the USA. She explained that she would have matured anywhere, as long as she pursued her doctoral studies in a location in Taiwan where she had to live apart from her family. But she had also wanted to find a place where her gender was not viewed as a handicap or a disqualifier. For that, she had to leave Taiwan:

> I thought [the USA] would be the place where I can be independent and can learn something and can do something away from tradition. Away from whatever was expected from us [women]. Like "a woman should not do this or that" kind of tradition.

In Chapter 6, I discuss more of the gendered migration decision-making and postmigration experiences of Asian women scientists, many of whom had experiences similar to Hsing-Yi.

There was the danger, however, of being given too much independence. A handful of interviewees reported having difficulties with absent or soon-to-be-retiring supervisors in their Western university who provided no support or mentoring. Interviewees also argued that particular country and university contexts had a culture of extreme independence that could be detrimental for international students who still needed some scaffolding to transition to a more independent research environment. Yoke Ling, who trained in the UK, spoke of how an excess of autonomy could backfire:

> In England, it's really laidback. Nothing is obligatory, you just have to go by yourself. But for Asian students, they often got lost. So I knew a lot of good Taiwanese students studying PhD in their different subject areas. But sometimes, I found they got lost because they could not work as independently as the supervisor was expecting them. [So] they just fail.

Interviewees like Yoke Ling recognized the need for effective mentoring even in an independent research environment, but they still concluded that the greater autonomy they experienced in the West had made them better scientists.

Importance Given to Rank and Seniority

Rank and seniority were also handled differently across Asian and Western labs, and this was somewhat linked to the differential emphasis on research independence discussed earlier. All of my interviewees were raised in societal cultures where the social (and scientific) norm was to respect elders and honor teachers. Moving to an environment where everyone was largely treated as equals in intellectual conversations was difficult to internalize at first. But over time, interviewees came to love such an environment.

Yong Kai described this experience as one more culture shock he had to adjust to. As he explained, "In Taiwan, young people, we are told just to listen. Don't try to speak from your mind. Just follow the order." Chao Xing, who had trained in both Scandinavia and Japan, offered a contrast between the cultures of these two places. While she saw the Scandinavian country where she completed her PhD as a "paradise for somebody who likes research," she found her Japanese postdoctoral lab to be very hierarchical:

> Japan is at the other extreme where hierarchy in interpersonal relationships can restrict you considerably. And in order to advance in terms of your academic pursuit, or even pushing through your ideas that you feel are right, this would require skills in dealing with people. So for those two years [I spent in Japan], I have to say I learnt more about that than research! Because it's really the extreme of how Asian societies are in terms of dealing with hierarchy.

Yiu Man, from Taiwan, was thrilled when he moved to the USA and experienced what he felt was a hierarchy-free environment focused on the common pursuit of knowledge: "You know, there's no leader, so to speak. I can freely talk to anybody just like he's my friend. ... It was just great." Yiu Man contrasted this environment with the situation in Taiwan where hierarchies based on age and rank mattered greatly:

> Taiwan is a Confucius country. You have to be respectful to the older people. You cannot just say [things] in very light-hearted or in a frank way. [Even people who] may train in the US, once they get back to Taiwan, it's different ... There is an invisible framework.

Yiu Man's point that people who are exposed to hierarchy-free training environments might temporarily internalize that value system, but then revert back to a hierarchical way of thinking when they return to their old organizational environment is an important one. There are two takeaways from Yiu Man's argument. The first is that Asian-born scientists are not intrinsically molded to a particular scientific culture, but can learn, adapt and adopt different scientific values and norms over time. The second is that for a set of values and preferences to scale up and become an organizational scientific culture there needs to be a sufficient number of people within the organization who espouse these values, reinforce and support each other, and then put these values into practice. Without a critical mass of practitioners, it is very hard for a shift in a scientific culture to take root, spread and become normative within a given organizational setting. As I discuss later in Chapter 8, as more and more Asian scientists return to work in Asia from various Western countries, their numbers are reaching the point where they can effect long-lasting cultural change in the research universities and institutes they have returned to.

I found that interviewees who returned to Asia often brought back with them an appreciation for egalitarianism in the workplace. They shared how they wanted to work in environments where they could speak their minds to their superiors, and where their students and staff would do likewise with them. They were willing to invest the time and energy to create a culture of equality within their labs, and they lobbied for similar change at a broader level in the organizations where they worked.

Attitudes towards Difference

Another dimension of scientific cultures that interviewees deemed important was the degree to which a scientific organization was open to individuals from different countries, cultures, educational backgrounds and identity traits. This dimension parallels the principle of universalism described by Robert Merton[16] whereby scientific ideas are judged on their merits, rather than the identity of the person proposing them. Interviewees regarded American universities as being more open to outsiders, in comparison to other Western countries' training environments. This applied both to the way that doctoral programs were structured in the USA versus Europe, as well as in terms of the learning environment within these programs.

Most American doctoral programs require students to complete approximately two years of master's-level coursework before commencing their independent doctoral research.[17] This allows doctoral trainees with varying educational backgrounds to build the necessary foundation of scientific knowledge and training before starting their independent research projects. In contrast, European doctoral training programs are usually shorter in duration and require almost immediate immersion in their supervisor's research project. This approach does not afford foreign students much chance to acclimatize to their new environment, or to fully develop their own research questions. Yiu Man, who trained in Taiwan and then a European country – and then reapplied for doctoral programs in the USA – argued that the American system was more welcoming to foreign students:

> In France or in other European countries like the UK or Germany ... it requires me to be with more background knowledge, you know? It's not good for someone like me who just came out of a system from another country, and immediately go into this European country system. It's too late.

Outside of these formal curricular structures, interviewees also spoke of how the American labs they worked in were more diverse in their student and staff makeup. As one Taiwanese scientist, Soon Huat, told me, "All the best people still go to the US. It's a multicultural nation. They don't care where you come from. As long as you are good, you can stay." Soon Huat recalled how his American doctoral lab had "people from France, from Austria, mainland China, Turkey, Bulgaria

---

[16] Merton 1973[1942].
[17] They also often have students rotate between different professors' labs so the students are exposed to different research questions, methodologies and work styles. Only after two to three rotations (each usually a few months long) is a student required to pair up with a professor who will become their doctoral supervisor.

[and] England." In contrast, the Taiwanese lab he trained in for his master's degree had only Taiwanese students. Soon Huat firmly believed that to be the "best lab," one had to accept people on the basis of their ideas, and not their identity. This is the crux of Merton's principle of universalism in science. Speaking about Taiwan, where he returned after many years in the USA, Soon Huat surmised: "We should become more multinational if we want to build hi-tech technology. You need to get the best people from outside, not just from local." To Soon Huat, Taiwan's social and cultural conservatism was holding back its scientific progress because foreigners viewed it as less accepting. Interestingly, he contrasted Taiwan with Hong Kong and Singapore, which he felt were two of the more open cultures in Asia and therefore more innovative.

Interviewees equated the openness of American institutions to foreign students and researchers with American norms of meritocracy. They interpreted their own admission into American labs as evidence that they had been judged on their intellectual promise rather than their ethnic, gender or national identity, and they appreciated this approach. Interviewees thus interpreted diverse labs as more competitive labs, because these labs were open to recruiting the best talent from anywhere in the world. According to Hock Peng, one of the greatest strengths of the American research system was how open it was:

> They are very aggressive about hiring internationals. You'd be surprised. I mean, it's really a melting pot. To their credit, and this is not meant as self-flattery, but I think that when they see talent they don't let [other things] get in the way.

Here, too, interviewees were quick to point out variation within the West in terms of different countries' relative openness to foreigners. While the USA was regarded as the gold standard in its acceptance of diversity, Europe was seen as much more closed. Vanya, who trained in four countries (India, the USA and two European countries), argued that the USA benefited from having a more accepting *and* more meritocratic culture than Europe:

> Well, I think the US has a certain energy that I didn't necessarily find as much in Europe. I found the US much more diverse in terms of the people that were there in the lab. There were people from all over the world. It was a genuine melting pot. And it really did have a sense of a meritocracy where, if you really,

really work hard, you'd make it to the top of those labs. ... I know that there is a glass ceiling [in the USA]; there's no question about it. But the US does still give you the sense of being a more inclusive scientific environment because it really is a melting pot, whereas in Europe that was not the case.

According to Vanya, the European country where she completed one of her postdoctoral fellowships was "still very national in a sense and it didn't have that feeling of diversity as much." While Vanya appreciated the work-life balance in that particular European country, she did not find its culture of scientific research as inclusive as the one she encountered in the USA. "As a scientific milieu, it wasn't one that I felt as happy or as included in as I did in the US," she shared.

Interviewees also repeatedly raised that American universities' scientific culture was more open to nonwhite foreigners. One Indian scientist, Aparna, now a full professor at an Ivy League university, spoke of traveling to Europe for conferences and being told by Asian students studying there about an entrenched bias against them in these European institutions: "When I go to Germany and I talk to ... foreign students, they all tell me that they don't think they have a future there. ... They said that it's the way things work. The jobs go to Germans." Aparna contrasted these accounts with her own experience pursuing doctoral training in the USA. She arrived fully expecting to be discriminated against, or at the very least somewhat marginalized, and was shocked when this did not happen:

> All the time, I was so surprised that I had people bending over backwards to help me out and so forth. Because it just didn't make sense to me. Even in India, foreigners are not treated as well as Indians. And so I just expected that I would always be an under-priority. And so it was mind-boggling to me that this didn't happen at all.

Aparna ascribed this open attitude to the fact that the USA views itself as an immigrant nation, as opposed to European countries that do not self-identify as immigrant countries.[18] Dipanjan, an Indian scientist who trained in multiple Western countries before accepting a job in Singapore, made similar observations about several European

---

[18] Alba and Foner 2015.

countries, and identified Singapore as having a more welcoming scientific culture compared to these countries:

> Science-wise and work-wise – Germany, France – they do really good work. No question. But still, they are kind of very closed too; they don't accept someone completely outside to enter into their system. . . . If I work there, even if I do go to work, I have to be like an outsider. Which is not at all the case in the US and Singapore.

Despite Dipanjan's compliments about Singapore, all four of my Asian case countries have an uneven record when it comes to their acceptance of diversity (whether along ethnic, religious, caste or nationality lines). The top universities in China, India and Taiwan are overwhelmingly staffed by native-born scientists drawn from specific segments of the native society. Even in Singapore, which has a high proportion of foreigners working in its scientific sector, some native-born ethnic groups (particularly Malays) are almost completely absent from the country's scientific domain. As I show in Chapter 6, women scientists experience significantly more difficulties pursuing successful careers in Asia, as compared to their male counterparts. Returning Asian scientists often came back with the goal of making their new labs as inclusive and international as the ones they left behind in the West, because they saw staff diversity as key to their lab's future research excellence. Later in this book, I discuss the challenges they encountered in these diversity and inclusion efforts.

Approach to Communication

The final cultural difference interviewees highlighted between their various training environments related to the different attitudes towards inward and outward communication. Every organization – small and large – has its own "information culture," which relates to the practices and attitudes surrounding the sharing of information within the organization and also the dissemination of information to outsiders.[19] In Western and particularly American universities, interviewees reported a strong institutional focus on communication skills and information sharing. This occurred through graded presentation-based assignments in class

---

[19] Choo et al. 2008; Widén and Hansen 2012.

settings, as well as through regularly scheduled meetings within lab and project teams. Such knowledge sharing was deemed essential for collaborative innovation and learning to occur. These communication moments also had a critical reflection component, as presenters were expected to answer questions from the group and defend their answers with evidence.

For East Asian interviewees who came from a non-English language environment, the switch to an English-language learning environment was already hard. The additional emphasis on oral presentations and discussions in the lab and classroom was one of the hardest and most immediate challenges they had to overcome, as it accentuated their preexisting difficulties with the English language. Yan Jiang felt that this emphasis on oral presentations was the main difference between how he was trained in the USA versus China:

> The presentation skills. It's not only what you know, but also you have to present it and let other people know. They always make us present again and again, and learn those kind of skills. That skill – people don't think it's important in China. Because I didn't have any of that kind of training when I was in China.

Yan Jiang described how during his early training in China, communication occurred between individual students and their professor through written examinations and assignments. Any work done in the lab was supposed to speak for itself through technical reports of the results. The idea that students could learn from each other, rather than from their professor, or that students needed to learn oral communication skills, in addition to written ones, was not considered.

Interviewees described having to attend weekly lab meetings when they first started attending their Western university. During these meetings, they were required to present their work-in-progress to their supervisor and also their lab mates. Interviewees frequently noted how intimidating but also how intellectually stimulating these initial group meetings were. This norm of regular and engaged internal communications within the team was a habit that many interviewees only picked up in the West. (By the mid-2000s, this practice was increasingly commonplace in Asian labs as well.) Xiao Ling, who completed her master's in China before moving to the USA for her PhD in the late 1990s, worked in a top national-level laboratory in China before her westward move. She spoke of how little lateral sharing of research occurred during her time in the Chinese lab:

> We don't have group meetings, so basically I just talk to my direct PhD students or the postdoc in the lab about my research progress. And to be honest, I really don't know what's going on. I don't even know what my project is about. I know I need to purify this protein, that's all. But what they're going to use this protein for, I have no idea. ... But here [in the USA], in the lab, we always have group meetings. So everyone will sit together and discuss their experimental results. We also have a journal club. I remember the first time I was asked to give a journal club or research talk. I was so nervous, I spent two weeks to prepare for it. And I did not have that chance when I was back in China.

In addition to intra-team information sharing, interviewees spoke of a greater emphasis in US institutions on the external sharing of their research with the broader public, which included funding agencies, the news media, and even laypersons, to communicate the importance of their work to nonexpert audiences. Part of this institutional emphasis stemmed from the earlier realization in Western societies that, for the public and the government to continue to fund scientific research, it behooved scientists to reach out to these groups and explain the relevance of science (and particularly basic science) to society. As Robert Merton pointed out more than fifty years ago, due to government budget cuts, Western scientists have had to accept that science is not set apart from society but is instead highly influenced by (and in turn, influences) events in the rest of the country. In contrast, this kind of reckoning had not yet happened during interviewees' early training in their Asian home country. As a result, there had been less institutional importance given previously in Asia to the need for scientists to communicate externally outside their expert circles about the value of their research.

Hock Peng spoke of how he learned the importance of this form of external communication while pursuing postdoctoral training in the USA. His fellowship at an Ivy League university was in the lab of a professor whom Hock Peng described as a natural "businessman." Until his fellowship, Hock Peng had thought that scientists just needed to focus on their experiments and benchwork, but his postdoctoral supervisor made him think about science in a whole new way:

> In my postdoctoral lab, we always have [nonscience] visitors coming in and then we have to give an impromptu presentation. Kind of like, you have to sell your ideas, tell them why it is so important. My boss is an extremely good businessman. Every single small idea, he can talk about it like it is the best idea in the world. I was very amazed by how he operates but actually, as he put it, "A lot of science – to be successful, it has to be run like a business." It's not just good research. Good research is one part, but being able to sell your science, being able to communicate why your idea is so important is equally important.

For Hock Peng, the importance of communicating his ideas to laypersons was a value he learned in the USA and took back with him to Asia, full of plans about how he was going to establish an open information-sharing culture within the new lab he was going to set up.

Conclusion

This chapter has demonstrated the varying scientific research environments and cultures that existed in the Asian and Western countries where my interviewees underwent training. The scientific cultures they were exposed to when they trained in the West had a significant impact on how they, as individual scientists, subsequently approached scientific knowledge and scientific research questions, worked with others in their own lab, and the relative importance they gave to questions of diversity, communication, critical thinking and egalitarianism in research settings. From their accounts of their training experiences, I identified seven constituent dimensions of any given scientific culture. While more can (and should) be done to create a more granular understanding of a research team or organization's scientific culture, the schematic presented in this chapter (see Figure 4.1) can hopefully be used to analyze the scientific cultures of other labs, universities and even countries in the future.

This chapter has also revealed some of the variation that exists within the West, but also within Asia, along each of the seven dimensions of scientific cultures. The scientific culture of elite American universities was seen as more ambitious and more open to foreigners compared to top European universities and research institutes. Both American and European universities were seen as having scientific cultures that encouraged an inquiry-driven approach to scientific learning and fostered a higher degree of autonomy and information sharing among junior scientists. Meanwhile, within Asia, Singapore was perceived to be the most multicultural and accepting of difference. In this manner, scientists shared with me the different ways of teaching and practicing science that they had encountered during their training, all the while pointing to the ideal scientific culture they wanted to work in and foster in their own labs and classrooms. Interviewees had specific ideas about the kind of scientific culture they saw as enabling the most impactful research, and the scientific culture they wanted to nurture in their places of work.

This leads me to the third and final takeaway of this chapter: that there is nothing intrinsically "Asian" about a particular scientific culture. The many years interviewees spent training in the West altered how they thought the social practice of scientific research "should" be taught, organized and managed. What this means is that there was nothing that hardwired my Asian interviewees to practice science in a particular way. Having been trained in multiple scientific cultures, they adopted and adapted those aspects of each culture they liked the most and what they saw as achievable within the broader organizational and societal cultures in which they were now embedded. As the brain circulations between Asian and Western countries increase, we should expect even more hybridization of scientific cultures to occur.

5 | *Return to the Future or the Past?*

> Evolution is not about competition. I'm telling you this because there's a reason why I came here [to Singapore]. Competition in a given population betters the population to a certain point, but competition is not the reason why new species are created. If you read carefully *On the Origin of Species* from Darwin, this point comes across very clearly. The mechanism for new species to be created is different. Competition betters a population along the same tracks; to create a *new* species, one of the factors is geographical isolation. You are cut off from the mainstream and you take your own path to evolution – that's how you create a species. Slowly, you become isolated reproductively. Slowly, something changes in your genes because you can't mate with the original population, you only mate with those guys who got away, and then you start producing other species.
>
> From a social point of view, for rudimentary ideas which might be off-the-wall, you require a degree of isolation. Because, if you are in the middle of all these people, ... you will just do what is approved by them. So because here [in Singapore], nobody knew me, and nobody knew Singapore, perfect. Let's go!

This was how Vivek, an Indian scientist, explained why he left Europe in the mid-1990s, rejecting the offer of a position at a well-established research institute to move to Asia, and specifically Singapore. At the time, Singapore did not have a reputation as a global research hub. One of Vivek's advisors was incredulous when Vivek shared that he was considering moving to Singapore. "Are you crazy?" his advisor exclaimed. "Why Singapore? Singapore is an old British colonial post. They do business, not science. It's suicide!"

Vivek himself did not know much about Singapore before being invited to fly out for a job talk. But once he met and talked with his future colleagues, he became intrigued by the creative possibilities of "a certain degree of isolation" and accepted the offer. "If you are surrounded by people who are very famous, your thinking gets overshadowed," he told me. "You always worry about what they think about your ideas. . . . And if somebody doesn't fit, they will tell you [that your ideas are] nonsense [and that] you're wasting your time. Sometimes a bit of isolation is important."

Contrast Vivek's renegade explanation for his return to Asia with that of Pei Chia, a Taiwanese physician scientist who completed her medical training in her home country before moving to the USA for a PhD. After several years in the United States, she returned to Taiwan to join a university in Taipei in the late 2000s. Pei Chia shared that return had been her plan all along:

> I decided to come back to Taiwan even before I went to the US. And those ideas were not changed during my PhD course. I think there are several reasons. One is for my family. Although my parents are healthy and I have many siblings, I still feel obliged to come back to Taiwan to take care of my parents.
>
> And then, I have a very strong tie with my relatives and my friends and my teachers in Taiwan. That's the second thing. I know many PhD students from Taiwan. If they went to the US during their twenties, a higher proportion of them stay in the US. But for people who went to the US in their thirties[, like me], usually they all came back to Taiwan.
>
> And my third reason to come back to Taiwan is that I completed my [medical] residency training in Taiwan. So I've got quite a strong personal relationship with many senior members in the hospital. I have many friends here, and I have my license to practice medicine in Taiwan, so that's important for me to make a good living and to have my research. If I stay in the US, even though I can do good genetic study, I do not have the connections to get help or get patient resources from other doctors. But in Taiwan, I personally have patient resources and my colleagues will refer patients to me for study. So in terms of the research resources, for me, it is much better to be in Taiwan than in the US.

Pei Chia's justification of her decision to return to Taiwan in the late 2000s highlights the power of the social (in the form of family, friends *and* professional networks), as opposed to Vivek's argument about the allure of social isolation from established research networks. A second difference between these two returnees is that, while Vivek "returned" to a part of Asia where he had never lived, Pei Chia was returning to her birth country. A third contrast between their return decisions had to do with the perceived risk of their respective moves back to Asia. Pei Chia's decision was not about risk-taking; in many ways, it was the safe choice for her. She was drawn by the prospect of greater job stability and promotion prospects in Asia, even if the job may not have paid as well as those in the USA.[1] For Vivek, however, turning down a job at a renowned European institute for one in Singapore involved a leap of faith.

How should we approach this study of return migration, when it can represent different things to different people at different points in time? The traditional approach taken by migration scholars assumes a finality in the act of return.[2] It is presumed to mark the end of an individual's journey and the culmination of their long-term mobility project. When viewed as an end point, return is often judged in dichotomous terms as an indication of either the success or failure of the earlier migration. Understanding migrants' reasons for return would then only involve studying their overseas experiences.

However, for scientists and other migrants who have in-built mobility expectations within their career course, return may be the *beginning* of a new chapter in their careers. This return may have been planned from the start of their initial out-migration (as in Pei Chia's case), or it might be a response to an unexpected opportunity or constraint (as with Vivek). Unlike the vast majority of Asian labor migrants on short-term contracts, Asian scientists (as high-skilled, in-demand migrants) may be exercising their *option* to return. In such cases, when unpacking the return decision, it is also important to consider what returnees imagine their future lives will be like back in Asia. As anthropologist Xiang Biao writes, in the early twenty-first century, a return to Asia may be "perceived as a 'return to the future' – in the rush ahead of global business and technology curves. Return is a project driven by enterprise rather than by nostalgia" (2013:2). I acknowledge Xiang Biao's point but I also argue that the return decision for Asian scientists involves a constant tacking back and

---

[1] Chen 2015.
[2] See Cassarino (2004) for alternative ways to theorize return migration.

forth between ideas of the past and the future versions of home. This argument is the inspiration behind this chapter's title: *Return to the Future or the Past?*

Of the 119 Asian scientists I interviewed, 52 had returned to work in their birth country while another 34 were working in an Asian country different from their birth country. The remaining 33 interviewees had chosen not to return to Asia.[3] This tripartite distribution of return decisions (no return vs. return home vs. return to another Asian country) allows me to dig into the factors that influenced my interviewees' varying decisions.[4] Their accounts reveal that the decision-making behind return is more complicated than it was for the initial emigration to the West (detailed in Chapter 3). While interviewees spoke of the process of going to the West as a "train" they had to get on, they had been much more conflicted about whether or not to return to Asia. This was partly because of the different life stages they were at when pondering these two contrasting migration decisions. The optimism of their youth led my interviewees to ask fewer questions and take more risks when they first left for the West. In contrast, the return question arrived at a point in their life course when they were almost always the head or cohead of a nuclear family unit, often sandwiched between aging parents in one part of the world and young children and a spouse in another part.

I organize the factors which influenced interviewees' return decision along three axes of influence – integration, obligation and ambition. These three axes emerged as the most important criteria raised by interviewees in their justifications for their return or nonreturn:

(1) *Integration* refers to the social exclusion or cultural belonging that my interviewees experienced in their Western host country, compared to what they imagined they would feel back in Asia.

---

[3] A handful of interviewees first returned to their birth country and then moved to another Asian country, which is where I interviewed them. Others moved in reverse – working for a few years in a nonnative Asian country and then back home. An even smaller number moved back to Asia, and then left to "return" to the West. Again, these multinational migrations (Paul and Yeoh 2020) speak to the increasing diversification of the Asian scientist migration system.

[4] The distribution of return decisions within my interviewee sample varies from the return distributions within the broader population of Western-trained Asian scientists, but is still indicative of long-term trends of increased return to Asia.

(2) *Obligation* refers to how interviewees attempted to balance their sense of responsibility to different generations of their family: their parents (often still in Asia) versus their spouse and their children (in the West).
(3) *Ambition* references their particular scientific and professional goals for themselves as individuals versus as citizens, and where they thought they could have the greatest impact and/or career success.

Through my interviews, I found that these axes of influence are largely independent of each other. An Asian scientist's position on one axis may encourage them to return to Asia, while their position on another axis may push them to stay in the West. For instance, a Chinese scientist in the USA may feel increasing pressure from their aging parents to return to China, but they may also feel more American than Chinese after spending a decade in the West. Interviewees also reported shifts in their position on each axis over time, depending upon their particular life stage, career stage, and even the developmental stage of various Asian countries vis-à-vis the Western country where they lived. As a result, the desire to return to Asia could be stronger or weaker at different points in an Asian scientist's life.

In the past, the polarities of these three axes were more aligned on an East-West continuum. It used to be the case that feelings of social and cultural integration in the Western country where they lived, a sense of obligation towards younger family members, and an individually oriented scientific ambition encouraged Asian scientists to remain in the West after their training (see Figure 5.1). In contrast, feelings of social exclusion from their particular Western society, a strong degree of filial obligation to older family members, and a sense of duty to aid their home country's scientific and developmental ambitions encouraged scientists to return to Asia. The orientation of two of these axes – a scientist's degree of Western integration and the orientation of their sense of familial obligation – have remained largely the same in how they influence the return decision. However, this chapter shows how the polarity of Asian scientists' ambitions has changed significantly in recent years. Ambitious Asian scientists may no longer feel they need to make a choice between their desire to produce cutting-edge science and their aspiration to help their nation's scientific development. Instead, what I found from many of my

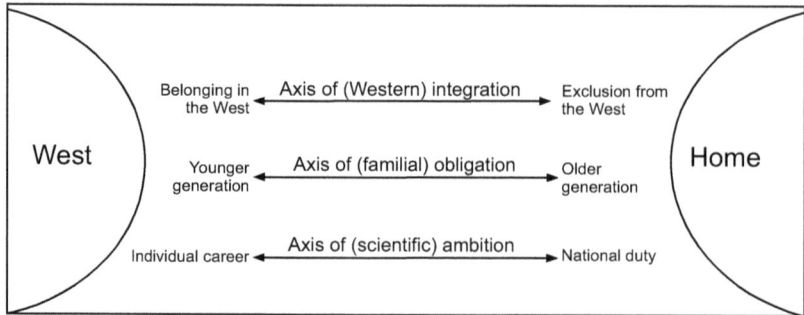

Figure 5.1 Traditional polarity of axes of influence around the return decision for Asian scientists in the West
Note: the polarity of the axis of scientific ambition no longer neatly aligns along West-Asia lines.

returnees' decision logic was that, starting in the late 2000s, the scientific appeal of various Asian countries, as well as specific Asian research universities and institutes, had increased to the point that ambitious Asian scientists now felt that a return to Asia held relatively more potentialities than problems.

As with Chapter 3, this chapter deconstructs the decision to return to Asia into three subquestions. In addition to the decision to return or not, I discuss interviewees' decisions about the *timing* of their return, and also the question of *where* to return. Linked to this last point, I use this chapter to introduce the alternative return migration arrangements some of my interviewees adopted as a compromise between the return-versus-no-return binary. These alternatives included leaving the West and moving to an Asian country other than their birth country, as well as establishing a transnational, split-household arrangement where one spouse returned to Asia and the other did not. These alternative arrangements highlight the growing complexity of migration options being considered by Asian scientists and the new patterns of brain circulation within the Asian scientist migration system.[5] In contrast with Chapter 3, this chapter focuses on the middle and tail end of Asian scientists' career courses, not the beginning. I explore how return migration may foster either "cracks" or "jumps" in my interviewees' timelines.[6] Before I discuss these findings, however, I detail the existing

[5] See Figure 1.3.
[6] Nowotny 2017.

literature on voluntary return migration among the highly skilled, to highlight some of the factors other scholars have identified as shaping return deliberations.

Return Migration among the Highly Skilled

Recent research has challenged the dichotomous interpretation of return migration in terms of economic failure or success.[7] These studies interrogate the factors that may influence return when there is a choice being exercised by the migrant.[8] Individual characteristics such as life stage, family background, educational background, degree of cultural assimilation overseas, and previous migration history have all been found to influence the likelihood of return migration.[9] Such decisions are also connected to the social networks and professional opportunities available in migrants' home countries.[10]

The presence of family back home is an oft-cited reason for voluntary return migration.[11] Part of this stems from migrants' sense of filial obligation towards their aging parents.[12] But desiring to be closer to family may also be linked to returnees' entry into a new life stage, such as parenthood or the dissolution of their marriage. Return migrants may choose to bring their children back to their home country not only to raise them in their traditional culture, but also to draw on childcare support from family networks.[13] However, women migrants are often more reluctant to return, especially if they enjoy more equitable status with male household members overseas and worry about reverting to a secondary status back in their country of birth.[14] This was the case for several of my female interviewees who resisted return when they could, as I discuss in this chapter and the next.[15]

[7] Hercog and Siddiqui 2014; Kōu and Bailey 2014; Wang, Tang and Li 2014; Harvey 2009; Chen 2015; Constant and Massey 2002; Xiang 2013.
[8] These studies of voluntary return exclude return flows of repatriated asylum seekers, short-term labor migrants whose contracts have come to an end, and irregular migrants who are deported.
[9] Hercog and Siddiqui 2014; Harvey 2009; Ammassari 2009.
[10] Wang, Tang and Li 2014; Harvey 2009.
[11] Gibson and McKenzie 2011.
[12] Kōu and Bailey 2014; Harvey 2009.
[13] Bhatt 2018; Upadhya 2013; Laoire 2008.
[14] Guarnizo 1997; Pessar 1996; Hondagneu-Sotelo 1994.
[15] I highlight such stories in Chapter 6. These women scientists' resistance to return points to the need for migration scholars to problematize our engagement with

A desire for cultural, alongside social, integration can also be a factor driving return. A study of highly skilled Indian migrants in Europe revealed that the inability to integrate, such as a lack of fluency in the host society's language, was correlated with a migrant's desire to return.[16] Several of the Chinese and Taiwanese scientists I interviewed raised this issue of language fluency, but not my Indian and Singaporean scientists who were all educated in English-medium schools and were fluent in English. British-born and Indian-born scientists in Boston likewise cite the culture and lifestyle in their respective birth countries as being a significant consideration if they decide to return.[17] Similarly, Chinese academic returnees emphasize family ties but also a sense of cultural belonging as key motivations behind their return.[18] Meanwhile, institutional factors such as labor laws and discriminatory hiring and promotion structures that prevent high-skilled migrants from feeling accepted in their overseas workplaces can also prompt return.[19]

While these studies may give the impression of a halcyon life in the birth country awaiting the returnee, this is rarely the case.[20] Migrants who spend a considerable amount of time overseas may experience a sense of alienation upon return.[21] Sociologist David Fitzgerald puts forward the idea of "dissimilation" to describe the ways in which migrants change and grow apart from the culture of their home country, making reintegration after return difficult.[22] Even the *fear* of experiencing dissimilation at social, cultural or psychological levels can discourage return. High-skilled migrants may be reluctant to return to a town or city that lacks like-minded individuals of a similar background. For this reason, the question of where to return becomes pertinent as the potential return destination can help (or hinder) the reentry and reassimilation process.

---

ideas of "home" and treat returnees' country of birth as almost akin to a foreign country that the return migrant must (re-)acclimatize to. In particular, the next chapter makes a call for more gendered analyses of return decisions and experiences (see also Ackers 2004).

[16] Hercog and Siddiqui 2014.
[17] Harvey 2009.
[18] Chen 2015.
[19] Varrel 2011; Raghuram 2006; Creese, Dyck and McLaren 2011.
[20] Ammassari 2009; Guarnizo 1997; Fitzgerald 2013.
[21] Bovenkerk 1974. I dig deeper into this idea in Chapter 8.
[22] Fitzgerald 2013.

The questions surrounding the return destination decision occur at a relatively granular level, with potential returnees making comparisons between individual cities, universities and institutes, rather than simply at the country level. Asian scientists in the West ask specific questions about what their lifestyle and work experiences will be like in each location under consideration – "Will I be able to recruit the right kind of staff for my lab?", "Will my children like their new teachers and classmates?", "Will I get along with my new colleagues?" – before deciding if they should return or not.

Even as a migrant's values, priorities and outlook may change during their time overseas, influencing their return decision, their birth country likely also changes. This change within the home country can make it a more or less attractive return destination for the migrant. When return happens in large numbers during a specific period in the history of a country, it is usually linked to a significant change in the developmental, social and political context of the country.[23] For example, Indian software professionals began returning in noticeable numbers to India in the 1990s after the Indian government began liberalizing the economy and allowing significant FDI into the country. Likewise, Taiwanese engineers started returning to Taiwan in larger numbers in the 1980s, with the end of martial law in 1987 and the increasing globalization of Taiwan's industrial sector.[24] Chinese entrepreneurs began returning and forming transnational households linking North America with mainland China in the 1990s and 2000s, after China liberalized its economy, joined the World Trade Organization, and began investing in high-skilled industries.[25] Likewise, the vast majority of the scientists I interviewed came back to Asia in the 2000s and later (see Table 5.1), when the changes to the tertiary education sector and scientific research system in each of my four Asian case countries began to pick up steam – as detailed in Chapter 2.

In all four of these countries, part of the draw of home was returnees' *re*conceptualization of their country of birth. In these migrants' minds, a new country now existed or could come to exist, if they returned. Along similar lines, anthropologist Carol Upadhya argues that the software professionals who returned to India during

---

[23] In this sense, a returnee can be viewed as akin to an exile who returns when conditions in the home country change.
[24] Saxenian 2005.
[25] Ley and Kobayashi 2005; Ong 1999.

Table 5.1 *Number of returnees, by country of birth and decade of return*

Country of birth	Number of returnees by decade of return			Total number of returnees
	1990s and earlier	2000s	2010s*	
China	5	9	13	27
India	3	14	11	28
Taiwan	6	11	3	20
Singapore	1	3	5	9
Total	15 (18%)	37 (44%)	32 (38%)	84

Note: returnees from South Korea and Malaysia are not included in this table.
* I conducted my research midway through the 2010s, which helps to explain the lower return numbers for that decade given that it was closer to only a half-decade.

the early 2000s displayed a neonationalism when they spoke of helping "to build the 'new India' by returning to share their wealth, knowledge, and entrepreneurial skills" (2013:142). Do returning Asian scientists similarly see their role as energizing the scientific progress of their country of birth? Or do they instead see themselves as pursuing projects of individual (rather than national) ambition? These were some of the questions I discussed with my interviewees when we came to the topic of return. Being Asian and a "halfway-returnee" myself, put me in an advantageous position to ask these questions, leading to frank and sometimes emotional conversations about interviewees' fears and hopes over their return decision and their successes and disappointments after return.

Of the three axes of influence shaping their return decision, interviewees confirmed that only one of these axes had changed radically in the last twenty years. A key contribution of this chapter is its demonstration that, between integration, obligation and ambition, it was interviewees' ambition (and how they viewed Asia in relation to their ambition) that was the key driver behind the recent spike in the return migrations to Asia. Scientists like Pei Chia who felt better integrated in Asia, and other scientists who felt obligated to look after aging parents, had always leaned towards returning, and this pull to Asia has not changed substantially over the years. The difference now is that Asian scientists like Vivek, who had ambitions to pursue field-changing scientific research, have also begun seriously considering return. As I demonstrate in this chapter, Asia used to represent the past for Asian scientists who associated the West with modernity, career advancement and the frontlines of scientific research. But increasingly, Asian scientists see Asia as the future.

Return Statistics

Before discussing my interviewees' return decision-making process, I share some statistics about the patterns of return they undertook. On average, returnees spent 11.8 years in the West before their move back to Asia, and the majority lived in the West between five to fifteen years. Interviewees on the short end of this range completed their PhD in Asia and then moved to the West for one postdoctoral training stint before returning home, while interviewees on the long end completed both doctoral and postdoctoral training in the West before returning. The average time spent in the West was remarkably consistent across the four case countries, with the average time spent in the West being 10.9 years for interviewees in China, 11.3 years for interviewees in India, 11.4 years for those in Singapore, and 11.9 years for interviewees in Taiwan. Five of the 86 returnees I interviewed had no option but to return as they were on government scholarships that required them to return after completing their overseas training. Two of these scholars were Singaporean and three were Taiwanese.

Regarding the decade of return, there was some variation by birth country (see Table 5.1). Though I conducted my interviews between 2014 and 2016, only halfway through the decade, a relatively high proportion of returnees (38 percent) had come back to Asia since 2010. Almost half of my Chinese returnees had returned after 2010. A significant proportion of them came back under the auspices of China's Hundred Talents and Thousand Talents Programs, which provided returnees with a full professorship and generous salary, benefits, and research support after a minimum number of years of research experience overseas. Though their numbers were significantly lower than their Chinese counterparts, more than half of the Singaporean returnees I interviewed also came back to Asia in the 2010s. In contrast, half of the Indian and Taiwanese returnees returned in the 2000s.

Table 5.2 *Number of returnees, by country of birth and country of interview*

Country of birth	Country of interview				Total returnees by country of birth
	China	India	Taiwan	Singapore	
China	13	0	1	13	27
India	0	15	0	13	28
Taiwan	0	0	20	0	20
Singapore	0	1	1	7	9
Total returnees by country of interview	13	16	22	33	84

Note: returnees from South Korea and Malaysia are not included in this table.

Digging deeper into the question of where people returned, the downward diagonal in Table 5.2 shows that my interviewees in China, India and Taiwan were overwhelmingly nationals of China, India and Taiwan respectively. However, my thirty-three interviewees in Singapore represented a mix of nationalities. The majority were not Singaporean, but instead Chinese and Indian nationals. Their presence in Singapore speaks to the rise of Singapore as a regional and global research hub over the last couple of decades, and the country's appeal to nonnative Asian (and non-Asian) scientists because of its proximity to the rest of Asia, Global North standards of living, extensive government investment in R&D initiatives, use of English as its language of education and research, and its reputation for political and socioeconomic stability.[26]

Deciding to Return

Having provided some overview statistics about my returnee interviewees, I now turn to their qualitative justifications for their return, organized along their three main axes of influence: integration, obligation and ambition. As in Chapter 4, which dealt with the diversity of scientific cultures that exist in Western and Asian countries, this chapter also complicates the metacategories of the "West" and "Asia." In

[26] Ong 2016; Wong 2011. See also Chapter 2.

describing the relative appeal of different countries, interviewees rarely spoke in terms of the "West" versus "Asia." Instead, they referenced particular characteristics of individual countries, and even went down to the level of individual cities or research organizations.

What their accounts revealed was both expected and unexpected. As the literature predicts, interviewees who felt themselves less socially and culturally integrated in their overseas Western country were more likely to have returned to Asia. Similarly, scientists who expressed an overarching sense of filial obligation to personally care for aging parents chose to return home, if they could not convince their parents to emigrate to the West or if they were the only child. But what was surprising to me was how recent returnees' personal ambitions had also fostered a positive attitude towards the idea of return, swaying their decision-making in previously unheard-of ways. In the sections ahead, I discuss each of the three axes of influence affecting the return decision.

Integration

Within migration studies, significant research exists on the factors influencing the assimilation and integration of new arrivals in a country, and how that may influence their future socioeconomic mobility as well as their likelihood of permanent settlement overseas.[27] While the term "assimilation" is largely used to refer to the process by which new arrivals adopt the cultural norms, values and beliefs of their host society, "integration" is a broader concept emphasizing the ability of new arrivals to access the same level of opportunities – employment, political representation, education, housing, health and nondiscrimination – that similarly positioned native-born individuals in their host society enjoy. We tend to assume that highly skilled Asian immigrants to the West will find assimilation and integration relatively easy, thanks to their educational advantage. But their status as racial and ethnic minorities can still make full integration challenging.[28]

While recounting their experiences integrating into the Western countries where they had studied and worked, a handful of interviewees

---

[27] Alba and Foner 2015; Portes and Rumbaut 2001; Bauböck 2001; Alba and Nee 1997.
[28] Purkayastha 2005; Dhingra 2007.

shared that they had endured incidents of explicit and implicit racial discrimination while living in the West. But the majority insisted that their Western workplaces were largely absent of any racism. Almost all the male, English-educated Indian and Singaporean scientists I interviewed stressed that the scientific sector in the USA in particular was devoid of racism and instead highly meritocratic.[29] Shaun, a Singaporean microbiologist who trained in Europe and the United States, was effusive about what he saw as the lack of discrimination against Asians in the USA:

> It's probably the most diverse country in the world. So I don't really sense [any racism]. If anything, there's almost a reverse stereotype that Asians are too good. Like, too law-abiding, too nerdy, and studying too hard. So there is not really a negative stereotype.

Vinu, an Indian scientist who completed his PhD and postdoctoral training in the USA and currently works at a research institute on the US East Coast, emphasized that this culture of acceptance was unique to the USA:

> From my perspective, the US is the most welcoming. ... I'd say that that was always going to be a draw for me about the US. ... When you go to Europe, it's always enjoyable and the pace of life, the community style of doing things – I enjoy that very much. But I find the US far more accepting of differences across the board.

These comments echo what I shared from other scientists in Chapter 4. Later in his interview, Vinu stressed that the acceptance he experienced in the USA had been a deciding factor for him when it came to the question of where to settle down: "I do not want to go to a place, despite all its wonders, that is unable to bring itself to accept people who are from a different culture." Vinu was certain that he would have felt less welcome as a brown-skinned Indian man in Europe compared to the USA. Many interviewees echoed Vinu's contrasting opinions

---

[29] I do not agree with these interviewees' take on the absence of anti-Asian racism in the USA, but I acknowledge that this was their personal assessment based on their particular experiences as highly educated, English-speaking men working and living in liberal-leaning urban centers in the USA. The recent rise in anti-Asian hate crimes in the USA and other parts of Western Europe is an indicator of the continuing challenges for Asian immigrants in Western countries.

about American and European societies' openness to Asians.[30] One Chinese scientist, who pursued his doctoral and postdoctoral training in a northern European country, described how he had regularly bumped up against negative stereotypes about ethnic Asians during his time in Europe in the 1990s: "The local people – sometimes they don't understand. They feel that if you're from Asia, then you're nothing." This scientist argued that insular thinking and a lack of diversity had been widespread in European scientific circles at that time and that Asian scientists were sometimes mistreated as oddities. He recalled representing his European university at scientific conferences in Western Europe and being asked by other conference participants, "How did you come from [that part of Europe]? Are you a refugee or something?" Runchen, a Taiwanese scientist who studied in the UK, described her British experience similarly. She shared how happy she was to be back in Taiwan:

> I think [my research institution in Taiwan] is already the best place. We have everything we need and so why go outside [Taiwan]? And I have to say that I will still worry about race discrimination [outside]. ... Why we should suffer that? This is our own country, of course, and here we are all equal. But if we go outside, then always there's that kind of problem. So I cannot see why I should [go outside].

East Asian interviewees (in contrast with Singaporean and Indian interviewees) were more likely to have encountered racism from peers and professors during their time at Western universities. In a deeply emotional interview, Hong Joo, a South Korean scientist who pursued his PhD at a top university in the American Midwest in the early 2000s, spoke of being ostracized by some of his white lab mates:

> When I joined my adviser's group for my PhD, there's one Chinese student and four to five Western students. And three of those excluded me. I could feel it. And for the first six months,

---

[30] It should be noted that I conducted my interviews between 2014 and 2016 when Barack Obama was still the US president, and the country had a better reputation globally as a welcoming nation. But it is also the case that interviewees repeatedly stressed that the culture of their American universities was somewhat independent of the broader American culture. Some interviewees were living in US states that were not racially diverse, but in their universities and university towns or cities, they felt accepted.

> it was really, really hard for me. I still remember that time. And after six months, I went to see my adviser saying that I don't think I can make it here because of that [treatment]. It's hard to say but you can feel it, right? ... I cried at home. ... That was the only time I cried in front of my wife for my whole life.

Other East Asian interviewees did not experience as much overt discrimination in their doctoral or postdoctoral lab, but still felt like they did not truly belong in their Western environments. They saw this as the natural consequence of being an ethnic minority immigrant and opined that they would always feel somewhat marginalized in a Western country. Ting-Ting, a biostatistician from Taiwan who now works at a top American university, described feeling "lonely" at times as the only Asian in her subfield at her university: "You wonder why you're the only one. ... I go to study sessions, I review grant [applications] and I'm the only Asian there. Why is that? It feels like it shouldn't be." Likewise, a Chinese scientist who had eventually returned to China likened life in the USA to living in a hotel:

> Sometimes we feel, in the States, it's just like you stay in someone's house or stay in a hotel and it's cool. Like you're traveling for a month, you know, or a week, staying in a hotel. It is fun but if you think about a whole life staying there, sometimes you feel frustrated.

For some interviewees, these experiences of disjuncture and social distance in the West had coincided with a growing longing for their home country. Taiwanese bioscientist Rui spoke of cultural fit as being the deciding factor in her and her husband's decision to return to Taiwan, after they finished their postdoctoral training in the USA. For them, however, the question of "fitting in" referred, not just to their day-to-day living in the USA, but also to the culture of American universities and research institutes where English was the *lingua franca*:

> We all have the feeling that we will fit better in Taiwanese society. ... We are more comfortable with [other] Taiwanese. I think partly because of the English-speaking problem. Even though we could communicate, but we cannot really speak native English. And that kind of hinders us to be good friends or in the close circle in the institute [in America].

Deciding to Return 163

A parallel experience was recounted by Taiwanese scientist Chia-Ling, who spoke of how she had wanted to return to a setting where she had family nearby:

> The US, it's a big place. You don't have many friends, even though I joined a Christian fellowship. So I feel more comfortable to come back to a smaller place where I feel secure enough and my parents or my siblings are just nearby.

For scientists like Rui and Chia-Ling from non-English-speaking backgrounds, there was a strong appeal to being in an environment where their lack of fluency in English would no longer be such a handicap. They expressed this desire, even as they also expressed an appreciation of diversity in lab staffing.[31] Overall, however, the need to feel personally connected to, and embedded within, the broader fabric of the scientific and social communities where they lived was a key factor behind many East Asian interviewees' return decision.

Obligation

Returnees also revealed the deciding role that their sense of filial obligation to aging parents back in the home country played in their return decision. However, returnees shared that not all members of their immediate nuclear family held the same views on the question of return, with younger members (both the scientist's spouse and their children, if they had any) often preferring to stay in the West. Ram, an Indian biologist who went to the USA with his wife for postdoctoral training, tried to convince his parents to join him in the USA, arguing that he would earn more and that they would have access to better healthcare if he stayed overseas. According to Ram, his parents resisted this suggestion, arguing: "We don't know anybody there and what are we going to do there? For us, life is here [in India]." When he first left India, Ram had told his parents that he would spend between four to six years overseas before returning. He shared that "from the time four years got over, you know, the bells were ringing. 'What's going on? What's going on?' So there was pressure. And it was becoming bad."

---

[31] This contradiction between the cultural values interviewees espoused in Chapter 4 regarding the importance of diverse workplaces, and some returnees' simultaneous desire to be part of an ethnic majority proved challenging to reconcile, as I discuss in Chapter 8.

Eventually, the nagging became so draining that Ram gave up the fight and began applying to positions in India. After receiving a tenure-track offer from a top Indian university, Ram moved his household back to India over his wife's objections and his own misgivings.

Ram was not the only male scientist I interviewed who decided to return to Asia despite his wife's reluctance. Suresh, another Indian scientist, spoke of being torn between his parents, who wanted him home, and his Indian wife, who wanted to stay in the USA:

> I had always told myself I will go back to India and I had always told my parents that I would go back to India. . . . But then it's hanging. It's decision time, right? What am I going to do if I go back? So I wrestled with it a lot. And by then, I was married, so it's two people's decision. I had huge arguments also with my wife who didn't – I mean, no Indian woman wants to go back to India! I am sorry! [*Suresh laughs*]
>
> She said, "No way are we going back to India."
>
> I said, "No, we have to go back and it's good for this reason, that reason, and so on." So I said, "Okay, we are going back to India." But what am I going to do there?

Suresh ended up "compromising" with his wife by moving the family to Singapore (where he had a generous job offer waiting), rather than back to India, but this move still benefited him much more than it did his wife.[32]

Decisions were easier when both husband and wife were in agreement over their destination preferences. Xing Fu, a Chinese scientist I interviewed in Beijing, told me that many young US-trained Chinese male scientists want to return to China because they deem it to be better for their careers, but most Chinese wives want to stay in the USA because they prefer their family's lifestyle there. Xing Fu joked that he was lucky because his wife (who finished her PhD in the USA in the same field and at roughly the same time as him) had also wanted to return to China to look after her parents:

> Lots of the wives, they tend to want to stay in the US for many different reasons. But the males, they have really strong will to come back. So one of the reasons I can come back is just my wife. [Because] she also wanted to come back. Her parents live in

---

[32] I talk about the "gender compromises" that Asian women scientists make in their lives in much more detail in the next chapter.

Beijing and she's the only daughter of their family. My parents and her parents have been in the States for like half of the year. But they feel really bored in the States. So they won't go there [to settle down]. And we're thinking about later. They are aging and they need someone to take care of them later.

The tension over staying in the West versus returning to Asia was intricately tied to interviewees' sense of relative responsibility to older versus younger generations within their family unit. Taiwanese scientist Yiu Man framed his decision to return as a by-product of the Confucian culture of East Asian societies which inculcated the importance of filial piety in sons and daughters from an early age. While he enjoyed his life in the USA, he felt increasingly guilty about being the only member of his family not living in Taiwan. A death in the family finally instigated his return:

When in 2008 my father-in-law suddenly passed away, that kind of gave us a warning sign that everybody, every other important person in my life, could be gone in no time. And it would be a very true possibility that I may end up coming back to Taiwan always for a funeral or something. That's the thought that hit me at that time. ... When this thing happened, I suddenly realized I am a selfish son. And so I decided to actively call up my professors [in Taiwan], my previous advisors, to see [if there are any opportunities in Taiwan].

On the opposite end of the familial responsibility spectrum, interviewees spoke of how their children were a key factor in their decision to *not* return. Among the returnees I interviewed, forty-four had at least one child at the time of their return to Asia, with the modal number of children at the time of return being one (see Table 5.3).

Not all returnees told me the ages of their children at the time of return, but for the forty interviewees who did, I was able to calculate the age distribution of their children. It was clear that returnees with children typically came back when their children were very young. As Table 5.4 shows, half of returnees' children were five and under at the time of return, and another third were between the ages of six and ten. For example, Soon Huat returned to Taiwan with his wife and toddler-aged son after completing his doctoral and postdoctoral training in the USA. His son rapidly adjusted to life back in Taiwan because he was so young, and when I interviewed Soon Huat, his son was a college

Table 5.3 *Distribution of returnees, by number of children at time of return*

Number of children at time of return	Number of returnees	% of returnees
0	42	49%
1	24	29%
2	17	20%
3	2	2%
4	1	1%
Total	86	101%*

* The percentage total is greater than 100 due to rounding up.

Table 5.4 *Distribution of children of returnees, by age range at time of return*

Age range of children at time of return	Number of children	Number of returnees with children in each age range*
1–5 years	32	24
6–10 years	20	18
11–18 years	2	2
19+ years	7	3
Total	61	47

* The number of returnees adds up to more than forty because some returnees had multiple children in various age ranges.

student in Taiwan, looking to start his own independent migration to the USA for graduate school.

After their children turned ten, however, there was a steep decline in returnees within my interview sample. There were only two returnees whose children were between eleven and eighteen at the time of return, and one of these returnees chose to adopt a split-household arrangement because he felt that it would be difficult for his teenage son to handle a switch in countries, schools and friend groups. The other returnee was a Chinese scientist who took up a job in Singapore and moved his high-school-aged son to an international high school in

Singapore because he deemed that moving the family back to China would have been too big a culture shock for his son.

Many of the nonreturnees I interviewed cited the fact that their children had grown up in a Western environment, lacked fluency in their mother tongue, and were already in the middle- or high-school system in their particular Western country – all of which made a transition back to Asia challenging. Nomuro, who was one half of a Japanese scientist-couple I interviewed, explained to me that, while there were many reasons why he and his wife were reluctant to return, their preteen daughter was the primary factor. Nomuro did not appreciate what he saw as the hierarchical and patriarchal culture of much of Japanese society and was not willing to impose that culture on his daughter:

> Our daughter, she was born in the US and then raised like an American. She cannot adapt to the Japanese culture. It's completely different. If she goes to Japan, she will have a lot of problems because she looks like a Japanese ... but she cannot speak Japanese well. Even though we actually try to teach her to speak Japanese but, for her, it's a foreign language.

Return ticked up again slightly for more senior Asian scientists whose children were college-aged or older. There were three returnees who had, between them, seven children aged nineteen and over at the time of return, and their children had not accompanied them when they moved back to Asia. These scientists had been more open to the idea of moving back, as returning would not disrupt their children's lives as much. However, their spouses sometimes resisted return at this late stage in their life and career courses, especially if they were comfortably ensconced in their own networks and careers in the West.[33] Typically, return was easier to contemplate if scientists' spouses were experiencing integration challenges in the West and preferred to live in Asia. But if their spouses were well ensconced in their Western host society, then alternative return arrangements needed to be considered. Several senior male scientists I interviewed shared that they had adopted a transnational split-household arrangement when they returned in their fifties or sixties, because their wives had not been willing to uproot themselves from their successful careers in the West. In this manner,

---

[33] I highlight some examples of wives refusing to follow their husbands back to Asia in the next chapter.

interviewees were constantly negotiating between their sense of obligation towards different members of their family, each with their own location preferences.

Ambition

The final factor influencing the return decision is Asian scientists' ambitions. My interviewees were highly ambitious individuals, but the focus of their ambitions varied significantly. While some wanted to revolutionize their respective subfield and burnish their personal research credentials, others wanted to improve the state of scientific research in their birth country and saw their science as a tool for national development and progress. Scientists in the latter group were more likely to want to return home, even at some cost to their standing in the global scientific field and even if it meant a loss in net income. In addition, they were more willing to put up with the inefficiencies and contradictions in the scientific research environments they returned to in Asia. In contrast, the former group of scientists sought out organizations (and, by extension, countries) where they could access the best research environments to achieve their goals.

Several interviewees who were already tenured professors in American universities by the time I first met them, indicated that under-resourced research systems in their Asian birth country had been the key reason why they had ruled out return when they were younger. Uday, a tenured professor at a private American university, spoke of how he had rejected the idea of returning to India when he finished his PhD in the USA in the late 1990s:

> The only good [institutions in India] that I was really aware of were TIFR [the Tata Institute of Fundamental Research] and CCMB, which is the Centre for Cellular and Molecular Biology in Hyderabad. But at that point in time, even those institutions were still way behind American academic institutions. I decided to stay over here [in the USA] because, at that point in time, the project I had worked on had really flowered and I didn't want to abandon it when I really saw quite remarkable potential in those projects.

Uday believed that returning to India when he had tenure-track offers from several top American universities would be foolhardy. The same

Deciding to Return 169

unwillingness to deal with research roadblocks led another Indian scientist, Sai, to stay in the USA after his postdoctoral training ended in the early 2000s:

> I don't want to say that, in India, I won't be able to produce. For instance, NCBS [the National Center for Biological Sciences] is really great. But the thing is that it may be a bit harder. ... But it just so happens that things were rolling really well here [in the USA] for me. So why should I stop it and take a risk and go there?

When I interviewed him in 2016, Sai noted that the state of the scientific field in India had improved significantly since the early 2000s when he was considering whether or not to return. But when he had been considering return at the end of his postdoctoral training in the early 2000s, there had been strong reasons for staying in the USA.

As I described in Chapter 2, in the last two decades, all four of my Asian case countries – as well as other Asian countries like Japan and South Korea – have invested a significant proportion of their national budgets into enhancing their scientific research systems: establishing new national research institutes with access to generous block funding, restructuring their top national universities to make them more research-focused, and upgrading their research infrastructures. Richard, the scientist who moved back to Singapore from the UK, explained that the funding he was offered in his home country was more than he could have ever received in Europe:

> I'm able to do levels of research here [in Singapore] that I was not able to do in [my UK university]. [My subfield] uses a technology that is not cheap and, in the UK, it needs to be rationed. And because places like [my former university] have so many important people that have been in the place for a long time, they get first dibs. There's a pecking order. And in the place [in the UK] where I was going, it even shut down because they ran out of funding. Even a place like [my former UK university] has to align their resources. It's not a bottomless pit. They do have a lot of money, it's true, but here [Singapore] has far more.

Richard's decision to return to Singapore in the mid-2010s highlights the recent shift in the global scientific terrain that allows ambitious Asian scientists to pursue their professional dreams in Asia itself, rather than only in the West.

Chinese scientists at CAS also spoke to me of the greater volume of start-up grants, the larger laboratories, and the higher number of support staff they could hire in China under programs like the Hundred Talents Program, as compared to what they could do in the USA. (I explore more of these research perks in Chapter 7.) It did not hurt that, when they returned to China, these young scientists were also automatically given the title of full professor (with no attendant worries about getting tenure), rather than starting out as assistant professors as they would have in the USA. Xing Xing, a microbiologist who returned to China under the auspices of the Hundred Talents Program, identified the relative strengths of working in China:

> I have machinery in my lab now that I don't think most young PIs can get in the USA. I have ten persons in my lab now but if I work in the USA, maybe I have only two postdocs, one technician. That's like the most [I could have in the USA]. And maybe like a really small space.

Still, Xing Xing had not been blind to the ongoing disadvantages of working in China:

> It's relatively harder to do research here because, when you need some mice [for example], you have to order and they could take five weeks to get it [and] it costs like a couple thousand dollars. It's way more expensive here. And, for a reagent, we would probably have to pay more than 50 percent or maybe double [US prices] some time.

However, Xing Xing's start-up grant in China was large enough to absorb these increased costs. And because he could then afford to devote less time to grant writing, he had determined that it was still worthwhile to return to China. Uday, introduced earlier, shared that many Chinese postdoctoral fellows in his US department had returned to China in recent years because they followed the same logic. Uday was frank about which Asian universities he thought were now on par with respected public research universities in the West:[34]

[34] There was widespread acknowledgement among my interviewees that the very top private universities in the USA were still in a league of their own, with no real competition from top Asian universities for now. Returnees in Singapore, Taiwan and China instead made comparisons between the public universities they worked at in Asia and top public universities in the USA. The public University of California (UC) system, which is comprised of ten campuses

The University of Tokyo in Japan. I think some really good stuff is happening in Singapore as well. But, in addition to these, there are places in China. ... Tsinghua University, for instance. Five years ago, [before 2010,] if you had said Tsinghua University, I would have scratched my head and said, "What?" But now, it's talked about a lot. I know people personally at [my university] who have gotten positions at premier academic institutions in China and they've decided to go back. This is a sea change from when, you know, folks from China would come over here and then they would stay.

Many interviewees – especially those who had returned to Asia – echoed Uday, speaking of the growing capacity of select Asian universities and institutes to support cutting-edge research in the biological sciences. At the same time, returning scientists in China, India and Taiwan also called forth nationalist ideas to frame their motivations for returning. Personal ambition continued to be a key factor in the return decision, but interviewees who went back to their country of birth also spoke of how they were personally invested in their nation's scientific progress and development. Returnees in India and China in particular expressed a muscular desire to prove that they could produce research in their home institutions that was as good as anything coming out of the West. In this manner, both their individual and national pride were at stake when they justified their return.[35]

Smrithi, an Indian biologist, passionately defended her decision to return to India for her first tenure-track position after several years of postdoctoral training overseas:

> They say that when you are there [in the West], you can do so much better science there and you can't do it here [in India]. And I think years of listening to that made me so angry that I wanted to show that you could do this from here, right? There's nothing – I mean, there's no difference in the intellectual caliber between the best here and the best anywhere else in the world at

spread throughout the state of California, was the most frequent benchmark used. Interviewees would share how their particular Asian university was trying to move up the subjective rankings of the various UC campuses, with UC Berkeley considered the best.

[35] I acknowledge, however, that because I was conducting retrospective interviews, it is difficult to confirm if interviewees' nationalistic justifications were post hoc rationalizations of their return decision, or had in fact been part of their original decision logic.

the faculty level, at the student level. And I said, we *can* do it from here [in India]. There's nothing preventing us from doing it from here. . . . So that's why I think I wanted to come back. I just wanted to prove a point to people who are no longer here.

In both China and India, returnees spoke with pride about the rising quality of scientific output from the top universities in their respective country. But their arguments for returning were also wrapped up in their sense that their birth country *needed* them more than any Western country did. In other words, their ambition to pursue great science was linked to the question of what that science was for. Botanists from China and India spoke of wanting to create new strands of higher-yield, disease-resistant rice to bolster their nation's food security. Ecologists, meanwhile, were passionate about the possibilities of changing how their country approached environmental issues and interspecies interactions. Mary, an ecologist who moved to India with her husband, shared their reasoning:

> It was very clear to us by the time we were done with our PhD, that the US didn't need us. There are enough ecologists chipping away at grand and small questions all over the US. . . . But we can make a huge, huge difference elsewhere.

Mary acknowledged that India's research infrastructure was not yet on par with what was available in the West. But she believed that, from her tenure-track position at a top Indian university, she could help strengthen India's research systems and raise scientific standards within India's higher education sector for future generations of Indian scientists. Scientists like Mary, mainly in their early to mid-thirties when they first returned, saw themselves as part of a modernizing vanguard that could improve the state of their subfield in their country, and train and mentor its next generation of scientists.[36] (I dwell more on these motivations in Chapter 8.) By taking on PI positions in research institutions in their home country, they hoped to apply their overseas training to native fieldsites and populations, and to establish training bridges and knowledge exchanges between scientific communities in

---

[36] A similar return logic was espoused by Yuan T. Lee, Taiwan's first Nobel Prize laureate. An emeritus professor of chemistry at the University of California, Berkeley, he returned to serve as the president of Taiwan's Academia Sinica from 1994 to 2006.

the West and their home country.³⁷ Themes of sacrifice and duty were much more prevalent in these returnees' justifications. Anu spoke of her Indian family's multigenerational tradition of giving their lives to their country and saw herself as continuing that particular family tradition in her own special way:

> If you ask me, what's more important to me – science or doing science in India. Then I would say science and India are both equal. If science was the most important thing to me, then of course it's much harder to do science in India. I mean, science in the US moves at a much faster pace. Things are a heck of a lot easier. ... Plus, there's a critical mass there and it's a very meritocratic system. So if you simply want to do science, then it's tough actually to make a move back to India. But the science in India, it is actually equally important. So yeah, I would say both. The India component for me is actually quite important. I'm willing to take a lot more difficulties in terms of everyday managing of [my] scientific career and life and all of that, simply because it gives me the chance to do this in India.

Several Chinese returnees spoke with similar fervor, expressing a sense of national mission that at least partially overrode many of their frustrations with the work environment back in China. Liu Wei, a Chinese environmental scientist, felt compelled to work in his home country where the research priorities were very different from those in the USA:

> There are definitely trade-offs when I think about my career. In China, particularly for my field – ecology and environment science – there are lots of things we need to do and we have to do them *in* China. This is my home country and I feel, you know, like I have more responsibility here.

While a small number of my interviewees had been motivated to return to Asia even when the research infrastructures available back home were nowhere close to the levels available to them in the West, I do not want to stretch their example too far. It was clear from the overall return migration patterns in my sample that home-country scientific research systems still had to have crossed a minimum

---

³⁷ Chang (1992) observes similar patriotic reasons for return among Taiwanese scientists.

threshold before a critical mass of native scientists began contemplating return. In other words, for most Asian scientists, their sense of responsibility towards their nation only went so far. The role of the Asian state was thus pivotal in ensuring that the scientific research system in the home country was good enough to convince Asian scientists to even consider return.

In addition, it was clear that if an Asian scientist chose to return to an Asian country other than their birth country, no sense of pan-Asian patriotism drove their decision-making. Non-Singaporeans who chose to move to Singapore spoke of being motivated primarily by the state of scientific research systems (and particularly the greater access to research funding) in Singapore. For them, the value proposition of moving to Singapore was tied to the claim that it was on par with, if not superior to, Western research organizations. These scientists did not have any sense of national duty to moderate their frustrations if their new work environment failed to live up to their expectations. I posit that, partly for this reason, non-Singaporean halfway-returnees tended to be more critical of the scientific research systems and scientific cultures they encountered in Singapore, as compared to scientists who returned "home" to India, China and Taiwan.

Return When?

In addition to deciding whether or not to return, Asian scientists were also faced with the question of *when* to return. According to my interviewees, the most important rule of thumb to follow was to not return too early in one's career. The most common rank at which interviewees returned was at the assistant professor level or its equivalent (see Table 5.5), when their new job in Asia would also be their first independent research position, where they could build a lab of their own.

Returning earlier as a postdoctoral trainee was deemed unwise by most interviewees. A Singaporean bioscientist who moved back to Asia as a faculty member, after postdoctoral training in the USA, explained why moving back at the postdoctoral stage was ill-advised:

> If you came back to Asia too early, as postdocs, you kind of get swallowed up by the higher[-ranked] faculty. That's one thinking that we had. In the US system, we felt everyone was

Table 5.5 *Rank of returnees upon return, by country of return*

Country of return	Equivalent rank upon return (% of total)				Total number of returnees
	Postdoctoral trainee	Assistant professor	Associate professor	Full professor	
China	0	3	0	10*	13
India	1	14	0	0	15
Taiwan	3	15	3	1	22
Singapore	6	24	2	4	36
Total (%)	10 (12%)	56 (65%)	5 (6%)	15 (17%)	86

* These full professors in China were scientists returning under the Hundred Talents Program that gave full professorships to young scientists who had completed their postdoctoral training overseas. All ten returned to China in the 2000s and 2010s.

cooperative. Postdocs and faculty are kind of, more or less, on the same level in terms of the research. Faculty do respect them. They give quite a lot of independence to the postdocs and, in fact, postdocs are quite valuable because of their independence. They do their own stuff and all that.

But in Asia, ... there's still a little bit of a gap where, if you come back too early, it's a little less smooth.... There's a little bit of a tension there. When you are a postdoc here and then you become a faculty, you're still viewed as a postdoc by the people who were your faculty when you were a postdoc. The problem is that you're always viewed as a junior. So that was part of why we didn't try to come back too early. We wanted to gain more experience, build up our portfolio, and then when we come back, we would be in a better position. You kind of get a better deal and also have more independence.

From this scientist's point of view, the lack of respect for postdoctoral trainees was linked to what he considered a culture prevalent in many Asian countries, namely one that valorizes hierarchy and respect for seniority – an issue I first raised in Chapter 4 and will revisit in Chapter 8. Interviewees stressed that a young scientist would not be as respected if they completed their final training stage in their home country, as opposed to in the West. This difference in treatment also speaks to the ongoing valorization of Western-acquired human and cultural scientific capital, even as select Asian governments are trying to raise the profile of their own training systems. A US-trained Korean interviewee emphasized, "It's not a good path. . . . People believe that if you have the final education back in Korea, then the people who are in Korea may have less interest in you."

This systemic bias created a conundrum for returnees who wanted to hire talented postdoctoral trainees for their labs but could not find many because the best trainees still preferred postdoctoral positions in the West. An Indian scientist, who had completed his doctoral and postdoctoral training at Ivy League universities before returning to one of the most established scientific research institutes in India, lamented the shortage of high-quality postdoctoral candidates in India:

> So the quality of PhD students we get here is extremely good. But we basically get no postdocs. Basically zero. So although I have been here for almost three years now, I haven't had a single postdoc application. [And] it's unlikely to change unless there is a good postdoctoral program. In fact, I wouldn't encourage my own students to apply for postdocs in India, because starting from their salary, to research support, to everything, it is so bad that they might be potentially hurting their career. So unless they have other reasons for staying back, I don't think it's recommended at all.

Several of the returned scientists I interviewed in Asia had defaulted to encouraging their best undergraduate and master's students to pursue doctoral studies in Asia given the shortage of postdoctoral trainees. This was part of the reason why more aspiring Asian scientists are delaying their westward migrations until after their doctoral training, as they now have more opportunities to train under ambitious and high-achieving scientists in Asia itself.

Singapore was the Asian fieldsite where I interviewed the greatest number of returnees who had come back at the postdoctoral stage: six out of thirty-six returnees (see Table 5.5). Several of these returnees had been on government scholarships that required them to return at the postdoctoral stage of their careers, and most of them complained about this policy. To them, the Singapore government's approach was detrimental to their individual long-term success because they were not always assigned to a postdoctoral lab where they could pursue their particular research specialization. Inez, one such Singaporean scholarship holder, explained that, while she had always planned to return to

Singapore, she had wanted to return as an independent PI and not any earlier in her career:

> There's no big names in [my research] area in Singapore. So that makes it even worse because we know that the name of your [postdoc] boss matters a lot. And I think a lot of A*STAR scholars suffer from this. They work in a big shot's lab [in the West] as a PhD student. They do well; their bosses like them. They can place themselves in another big shot's lab and, in fact, my PhD supervisor he told me his advice is to join the biggest name you can convince to take you in for your postdoc. Because it's *that* important. The lineage matters. So if I stay in the States, I can, but instead, I have to go back to Singapore and join a nobody!

Setting aside Singapore, almost no returnees came back to my other three case countries at the postdoctoral stage. Returning at a more senior rank, when one was already established in the field, was also rare. Only fifteen of my eighty-four returnees had returned at the associate or full professor rank, but even this statistic is exaggerated because it includes the ten Chinese scientists who returned to China to the rank of full professor, even though they had just completed their postdoctoral training overseas (see Table 5.5). Once we set aside these returnees, the number of *truly* senior-rank returnees in my sample declines precipitously. Still, the small number of established returnees is understandable as the moving and switching costs for senior scientists are much greater and many of them have families firmly entrenched in the West by that time in their lives. The number of senior returnees was relatively high in Singapore because the country has instituted targeted programs to entice senior scientists to its shores, offering them generously endowed faculty positions with large research packages and travel stipends. One senior scientist, who was recruited from the USA to move to Singapore, shared how he was wooed by university leaders in Singapore:

> They systematically laid out a red carpet. I got a chance to set up an institute with resources that are fabulous, and they were very persistent and open and welcoming. I see in that sort of the US style, the anything-is-possible feeling. ... In terms of the leadership here, the aspiration is to be world-class.

In addition to a generous compensation package, this senior scientist was attracted to Singapore by the opportunity to lead a new center where he could put his ideas about how to pursue cutting-edge science in his subfield into immediate practice. Another senior scientist, who was recruited to restructure a university department, to shift it from a teaching-oriented unit to a research-focused one, spoke of returning so as to "have some impact, no matter how small it is, in terms of the science, in terms of the research in this region." In this manner, it is possible for senior scientists to be successfully wooed from the West by the chance to effect long-lasting and large-scale change and impact the state of scientific research in Asia.[38] But few senior scientists are usually willing to take on this challenge, and so the number of returning senior scientists remains low.

Return Where?

Alongside deciding whether or not to return to Asia, and when to return, the question of *where* to return in Asia needed to be settled. Interviewees raised a range of location characteristics that had to be available before they would ever consider return. On the personal and sociocultural side, these included their knowledge of the local language, the cost and standard of living in their new city or country, the quality of schools and childcare support (if they had children), the degree of proximity to extended family members, and the social and political freedoms they could enjoy in the location they were considering. On the research front, they considered the reputational standing of the institution they would join, the salary and benefits package, the overarching scientific research system, the national and internal grant funding situation, and the scientific culture in their particular institution.

But as Table 5.1 at the start of this chapter demonstrated, returning to Asia did not necessarily mean moving back to one's country of birth. Several interviewees preferred the idea of moving to an Asian country other than their country of birth, because the former country met more of their personal and professional criteria than their home country. Jian Kai, who was born in China and then trained in the USA, chose to move to Singapore while maintaining collaborative research projects in China. To him, this locational choice was ideal for him and

---

[38] In Chapter 8, I highlight how many of my interviewees in Asia spoke of their desire to impact the scientific field in their return country.

his family. Singapore was in Asia, but it offered all the lifestyle advantages of a Western country, while still being close enough to China that he could easily access the exciting research possibilities available to him in his birth country. He explained to me why he opted for this "halfway-return" arrangement:[39]

> After I decided to go back, I actually travelled [to China from the United States] a couple of times. I started to understand the system there and I know how difficult it is. I know what advantages that it can provide, but I also know what disadvantages are over there. So I decided, no, I will not go back to China full-time. I just think it's not a good model. The good model is to stay close enough, allowing you to work there, but you are not restricted by the system. So there are only two places I can look at [in that case]. I don't want to go to Korea, I don't want to go to Japan, because I don't want to learn another language. Learning English was already hard for me already! I just got English. I don't want to do that again!

Jian Kai deemed both Hong Kong and Singapore acceptable return locations and, after receiving offers from research institutes in both places, eventually accepted a PI position in Singapore. His wife, who had never enjoyed living in the USA after the 9/11 terrorist attacks, was thrilled with their new life in Singapore with its ethnic Chinese majority. Every few months, they would visit China to see family and friends, but they were always happy to return to Singapore, which they found somewhat boring but also more comfortable than China.

Several Indian scientists I interviewed felt the same about India, as Jian Kai did about China. They too had chosen Singapore as an alternative location in Asia to pursue their scientific research.[40] The advantages that Singapore offered included its status as "Asia-lite," meaning that it did not have any of the pollution, corruption, inefficiencies, or cultural and linguistic complexities that many other Asian countries were stereotyped as possessing, making it easier for non-Singaporeans to settle down there. The fact that it was already

---

[39] As mentioned in Chapter 1, I have adopted his term to formally define this alternative return arrangement.
[40] Hong Kong was another halfway-return option mentioned by several of my interviewees. Recently, however, Hong Kong's status as a viable halfway-return destination has come under question with the increasing incursions by China into its political space.

a multiethnic society also made it appealing to various other Asian nationalities.

Among the four Asian case countries I studied in depth, Singapore was the most successful in attracting nonnationals to its shores in significant numbers. At the same time, it was also the smallest country of the four, and therefore it was the country that relied most on nonnationals to boost its relative position within the global scientific field. Taiwan also recognizes the importance of attracting foreign scientists to its shores as a way to increase the diversity of its scientific human capital pool and encourage the cross-fertilization of ideas. There are lessons that Taiwan can learn from Singapore, in terms of how to make itself more appealing to nonnative Asian scientists. Similarly, there are lessons for Singapore from the other three countries in terms of how to placate nonnative scientists who might become frustrated or impatient with the local scientific research environment they encounter after return. I discuss these challenges in greater depth in Chapters 7 and 8.

Conclusion

The return migration of Asian scientists to Asia is a more complicated and emotionally fraught decision than their initial out-migration. Home-country governments often portray returning Asian scientists as national heroes, but my interviews revealed that the reality of these scientists' return decisions was more complex. From a theoretical perspective, this chapter shows that a variety of factors, revolving around three axes of influence – integration, obligation and ambition – contribute to scientists' return decision. Empirically, this chapter shows that, in the late 2000s and particularly after 2010, young and ambitious Asian scientists in the West became more open to the idea of return. The key driver behind this change was the improving scientific research systems in select Asian locations, which I discuss in Chapter 7.

Returnees to China and Singapore, in particular, were drawn by the large pool of research funds that were being put at their disposal, potentially allowing them to produce *more* research in Asia than if they had stayed in the West. Returnees to Taiwan were comforted by the fact that their institutions were well-resourced (though perhaps not at world-class levels) and that they would now be in coethnic environments that would make the social practice of scientific research – from grant writing to lab management to academic networking – more effortless. Returnees to India were the fewest in number, and all talked about how their return decision had been contingent on securing a place at one of the country's top research organizations and not just any institution back home. These scientists' professional ambition – an essential trait for a successful scientist – was thus a major factor in their return decision-making. Calling them selfless for their return decision ignores how they viewed their work and their particular return options. Few of my interviewees would have returned *only* out of a sense of duty, especially if their research would have suffered significantly upon their return. This was why interviewees who came from earlier cohorts of Asian migrant trainees had largely opted to stay on in the West rather than return – because the state of the scientific research systems in their home countries in the 1990s and earlier was nowhere close to what they could access in the West. And yet, as could be seen in the accounts from Mary, Anu and Liu Wei, the nationalist desire to help their countries' scientific advancement and development did play a part in returnees' multifaceted professional ambitions. And there was still a degree of sacrifice in the return decisions of some of these returnees, especially those from India and China who were returning to their home countries rather than some other Asian country.

Tied to the question of whether or not to return were equally important questions about *when* and *where* to return. In both of these domains, contradictions emerged that reveal the ongoing hierarchies within the global scientific field. Interviewees were unanimous that returning to Asia at the postdoctoral training moment was not ideal, even as they talked about their desire to recruit better-quality postdoctoral trainees for their labs in Asia. Likewise, returnees spoke at length about the importance of place in their return decisions, and some chose halfway-return locations like Singapore, which were better connected to Western scholarly networks, rather than returning to their birth country. However, even those who moved to a global city like Singapore acknowledged that, after their return, they felt somewhat outside the core networks within their subfields, as I discuss later in Chapter 7.

Importantly, none of my interviewees saw their return as the "end" of their career course. Instead, all saw return as the beginning of a new chapter in their scientific careers, often one that coincided with their transition from postdoctoral trainee to PI, with all of the new responsibilities and opportunities that this change in rank and status entailed. As a result, their return moment was filled with potentialities, rather than any finality in terms of success or failure in their scientific careers.

6 | Asian Women Scientists on the Move

There was always the implication there. If I was not so productive in a certain period, [my PhD advisor] would immediately comment that probably I'm not working hard enough. [Even though] I was working as hard as possible, but [to him] there's some sort of indication that the female is not strong enough to do that. ... My PhD advisor would tell things like, "Oh, you are not good enough as a scientist. You won't survive as a PI," or that kind of really negative remark. And so I came to the point where I thought, "I'm not good, I'm not good." ... Of course, I know I was a female but I never thought that's going to really influence my life or my decisions. But once I started my PhD, I started feeling acutely my gender.

The above quote comes from Sakuko, a Japanese bioscientist who almost gave up on a life of science because of how she was treated by her Japanese doctoral advisor. It was only after she left Japan to take up a postdoctoral fellowship in a top university in the USA that she rekindled her love of science and regained her confidence in her abilities as a scientist. Sakuko has since achieved tenure in the USA and received several awards within her field in both the USA and Japan. However, none of this would have happened had she stayed in Japan, which is notorious within scientific circles for its gender inequity issues.[1]

How do women scientists from other Asian countries fare in this regard? Like Sakuko, do they also find it easier to pursue their careers outside of Asia, and specifically in the West? Are there particular gendered challenges they experience even in the West? Are their brain circulation patterns in the Asian scientist migration system markedly different from those of their male compatriots? In order to answer these questions, this chapter considers the same social institutions and processes I explored in Chapters 3, 4 and 5, but now focused explicitly on

[1] Homma, Motohashi and Ohtsubo 2013; Ip 2011.

how they affect the international migrations and academic careers of the forty Asian women scientists I interviewed as part of this study. These women – from countries as diverse as China, India, Japan, Pakistan, Singapore and Taiwan – present a nuanced answer to these questions. This chapter outlines their gendered experiences pursuing a graduate education, postdoctoral training, and subsequently a career in science.

But at the same time, this chapter rejects the simplistic narrative that Asian societies or Asian men are overwhelmingly biased against women in science. Instead, most of the Asian women scientists I interviewed recounted similar experiences of class privilege and familial support that bolstered their early scientific ambitions. Almost all of them came from middle-class or professional backgrounds, and all spoke of the high levels of support they received from their parents and schoolteachers as they pursued their love of science.

However, when these women began shifting their status from "student of science" to "practitioner of science," they began encountering more gendered prejudice from their male peers and the broader scientific community in their home country. Other transition moments in their life course that also had a significant effect on their career course included their becoming wives, and later, mothers. Interviewees spoke of the familial and societal pressure placed on them to support their husband's career goals and carry a disproportionate share of the household responsibilities. Interviewees who returned to Asia also shared how these pressures ratcheted up upon their return.

In Chapter 4, I described the scientific culture shocks[2] that my interviewees experienced when they arrived in the West and were introduced to scientific cultures different from the ones they were exposed to during their early training years in Asia. This present chapter introduces the parallel idea of a "gender shock," which captures the experience of entering a social and symbolic space where the attitudes, norms and beliefs surrounding one's gender are unexpected and unfamiliar in either a positive or negative way. Upon entering this new space, the experience of gender shock heightens an individual's awareness of their gender, either temporarily or permanently. While culture shock is often used to describe the new cultural norms, attitudes

---

[2] Oberg 1960; Furnham and Bochner 1986; Zhou et al. 2008; Gaw 2000.

and values that are encountered when an individual moves to or visits a new country, gender shock can occur even in one's native country, simply by moving to a new social setting. Gender shock is a useful term to describe the self-reported gendered experiences of Asian women scientists as they moved through their life and career courses overseas and at home, encountering new social settings where their gender was viewed and treated differently from how it had been in previous social settings. Sakuko's description of "feeling acutely [her] gender" is the perfect distillation of this experience of gender shock.

As with Sakuko, some interviewees experienced a negative gender shock, where they felt inadequate or unwelcome in a particular scientific or educational space because of their gender identity. In other cases, they experienced a positive gender shock, where they suddenly felt that their gender was no longer a barrier to their learning or career advancement. A large segment of the women scientists I interviewed described experiencing one form of gender shock or the other, although the life stage and location where they experienced such a shock varied. In contrast, my male interviewees *never* spoke of experiencing a gender shock, positive or negative, revealing how much the scientific field remains a man's world. Several men spoke of experiencing scientific culture shocks upon arriving in the West (as I described in Chapter 4), but they never ascribed these experiences to their gender.

While gender shocks were reported by almost all the women scientists in my sample, my interviews also highlighted how significant variation exists across Asian countries in terms of attitudes towards women pursuing scientific careers. In the USA, I conducted interviews with women scientists from Asian countries as disparate as Japan and Pakistan. Even though these two countries sit at opposite ends of the development spectrum, women scientists from both these countries balked at the idea of return because of what they thought it would do to their career trajectories. These women had all experienced some gender discrimination in the West, but it paled in comparison to what they expected to encounter if they returned to work in their birth country. Meanwhile, returning interviewees in countries like China, Taiwan and Singapore were mainly concerned about intrafamily dynamics holding back their career progression, and did not describe the workplace culture in their home country as their biggest concern. Indian women scientists spoke about *both* the career and societal restrictions they had to endure upon return. Given this variation, I try not to make blanket statements about how gender "works" across all of Asia.

I term the corrective actions taken by Asian women scientists in response to these gendered social forces and gender shocks as "gender compromises."[3] I define a gender compromise as a settling for a less-than-preferred course of action in one domain of one's life (whether career, marriage, etc.) to satisfy gender-related pressures in the same or another life domain. Gender compromises are often predicated on the underlying informal gender contract that exists in a household and is regularly renegotiated by men and women in the face of social and economic changes, such as the entry of women into the formal labor force, women's increasing educational attainment levels, and (I argue) their experience of a gender shock.[4] This gender contract is expansive and often unspoken, and covers questions such as the normative division of domestic labor in the household, the relative importance of men and women's careers, and the locus of decision-making authority on household matters. Even as renegotiations of the gender contract in the household do occur, they rarely dismantle the underlying power inequality between the genders, such that women, and not men, are typically the ones that make gender compromises.[5] After experiencing a negative gender shock, Asian women scientists had to choose in which domain of their lives – family, career, education or location – they were willing to make concessions in order to adjust to their new gender reality. These ideas of gender shock and gender compromise provide a conceptual toolkit that captures how aspiring and existing Asian women scientists agentically navigate the growing educational and professional opportunities available to them, even as they deal with the ongoing patriarchal pressure to carry the majority of domestic responsibilities.[6]

---

[3] Choi and Peng 2016.
[4] Pfau-Effinger defines the gender contract as "the sociocultural consensus about the respective organisation of interaction between the sexes" (1994:1359). See also Bueskens 2018; Choi and Peng 2016; McDowell 2017.
[5] Pfau-Effinger 1994; Hochschild 1989; Bueskens 2018.
[6] Quah 2009. However, such gender compromises are not unique to Asian women scientists. There is extensive literature that shows how Western-born women scientists experience challenges too (see Ecklund and Lincoln 2016; Monosson 2011).

However, while the idea of gender compromise helps to partly explain the "leak" of qualified women scientists from STEM fields, it is important to note that women scientists' gender compromises do not always involve them giving up their career ambitions for the sake of family. It is also worth noting that women are not the only ones who make gender compromises. Several of the male scientists I interviewed made gender compromises too, but the conditions under which they made these compromises reveals the ongoing imbalance in power in most heterosexual relationships, as I show later. Before I detail these insights from my interviews, however, I provide a broad overview of some of the theories that have been developed to explain the status of women in science. Most of the work on this topic originates in the West,[7] and so I start there. I then introduce more recent case studies and statistics from my four Asian case countries – China, India, Singapore and Taiwan – before diving into my interviewees' own accounts.

Women in Science

As recently as the 1950s, there was still significant resistance in advanced Western countries towards women entering graduate programs in the sciences.[8] For aspiring female scientists of color, the odds were doubly stacked against them well into the 1970s.[9] Even today, some STEM subfields in the West (such as mathematics and physics) are considered more resistant to women than others.[10] Biology, in contrast, is often seen as a "soft" science and therefore more suitable for scientifically inclined women. In the life sciences, women doctorate recipients have overtaken men, with 55 percent of doctorates issued in the USA in 2016 being earned by women.[11] However, when it comes to hiring decisions in STEM fields, randomized control studies have found that male applicants are more likely to be viewed as more capable and subsequently hired compared to female applicants with identical qualifications.[12] As the sociologist and gender studies scholar Mary

---

[7] Watts 2007; Xie and Shauman 2003.
[8] Keller 2009.
[9] Sands 2009.
[10] Keller 2009.
[11] NSF 2018.
[12] Reuben, Sapienza and Zingales 2014; Moss-Racusin et al. 2012.

Frank Fox put it, "the idea of education, by itself, as emblematic of progress for women in science is questionable" (1999:453).

The concept of a "leaky pipeline" describes the steadily reducing proportion of women pursuing science subjects as we move from pretertiary to tertiary education levels, into the doctoral and postdoctoral stages, and then through the ranks of academia. In 2014, drawing on data from 112 countries, UNESCO reported that, while the worldwide proportion of women in science at the undergraduate and master's levels was higher than that of men, the proportion of women dropped close to 44 percent at the doctoral level, and then down to 29 percent at the researcher level.[13]

The leaky pipeline metaphor lends itself to a linear view of a scientist's career progression and to solutions that involve increasing the number of people entering the pipeline and reducing the number of people leaving the pipeline. Sociologists Yu Xie and Kimberlee Shauman criticize this pipeline approach for assuming a single, linear pathway to becoming a scientist and the framework's failure to recognize and make space for the various processes that feed into the leak of female scientists, particularly life-course events like marriage and childbirth.[14] Rather than using a pipeline approach to study the declining representation of women along the scientific career trajectory, they recommend investigating the "interactions of the multiple domains of an individual's life such as career and the family" (2003:4). Married scientists, for example, have to deal with their spouse's career ambitions in addition to their own. Research has shown that women scientists tend to give priority to their husband's career goals over their own. And if they have children, it is often mothers who undertake more of the childcare and domestic responsibilities. This can serve as a serious damper on their professional progression.[15] Despite their high levels of education, married female scientists may find themselves trapped within traditional gender roles that prevent them from committing as much time and attention as they would like (or need) into their careers. Lab- and field-based scientists, in particular, may face difficulties because of the long lab hours and travel required, which can take away from their family time. This can lead to speedbumps or

---

[13] UNESCO 2017, figure 11.
[14] Xie and Shauman 2003.
[15] Ecklund and Lincoln 2016; Monosson 2011.

cracks in their career timelines, as Helga Nowotny warns.[16] The late Finnish sociologist and STS scholar Veronica Stolte-Heiskanen eloquently described the conflict between "the biological clock, the domestic clock, and the research system clock" (1991:7). Women who want to start a family and still have some semblance of work-life balance may opt out of pursuing an ambitious scientific career trajectory, preferring to remain within the lower ranks of their research organizations or drop out of science altogether.[17]

Women in Science in Asia

Even though the number of Asian women scientists has increased over the last couple of decades, limited research exists on this population.[18] Nancy Ip (2011), a bioscientist at the Hong Kong University of Science and Technology, provides some useful statistics about the contemporary situation facing women scientists in East Asia. She notes that in the late twentieth century, East Asian women interested in a career in bioscience were often compelled to travel to the West to further their studies and seek work, because there were limited opportunities available in their own countries to find work as a woman scientist. Ip writes that in Japan, where a high degree of patriarchy prevails, women's representation in science is dismal. In 2008, only 12 percent of scientists in Japanese academia were women, with most passed over for promotion simply because of their gender.[19] The same pipeline leak noted in Western countries exists in Japan, with 27 percent of natural science undergraduates in 2009 being female, but only 13 percent of assistant professors in the sciences being women.[20] The difficulty of juggling work and family commitments, and the struggle to reenter the workforce after maternity leave are the two most cited reasons for women's "voluntary" absence from scientific fields. The norm that faculty members spend at least seventy hours each week on research is difficult for Japanese women with family commitments to meet.[21]

---

[16] Nowotny 2017.
[17] Ackers 2004.
[18] Campion and Shrum 2004.
[19] *Nature* 2008.
[20] Homma, Motohashi and Ohtsubo 2013:529. See also Lee (2010) and Marginson et al. (2013) for statistics on South Korea.
[21] Homma, Motohashi and Ohtsubo 2013:531.

Interestingly, Ip (2011) argues that in China, a country which has a long tradition of encouraging women to work, women scientists may be better placed to progress through the scientific career course.

In Taiwan, the privileging of science and technology in official state policy in the early decades of the country's independence made it easier for Taiwanese women to pursue STEM fields as a pathway to attain socioeconomic mobility; however, they did not necessarily enjoy significant career progression.[22] In the present day, wider career opportunities for women have led to fewer Taiwanese women pursuing science as a career. For Taiwanese women who do pursue a career in science, they tend to experience upward momentum in their careers only later in life, after their children have reached a more independent age.[23] Taiwanese husbands – even if they too are in the sciences – are rarely called upon to help with domestic matters. Women scientists in Taiwan typically manage this gender inequity by outsourcing as much of their household responsibilities as possible to older relatives or to paid, live-in domestic workers.[24] Other scholars have also noted the availability of cheap household help in several Asian countries as one way in which aspiring women scientists in Asia (as compared to in the West) manage to pursue productive research careers.[25]

The patrifocal and patriarchal nature of Indian society has been identified as the primary cause for the ongoing underrepresentation of women scientists at higher academic ranks and among the scientific leadership of Indian universities.[26] Sociologists Namrata Gupta and Arun K. Sharma (2002) report that 40 percent of the Indian women scientists they interviewed believed that their male colleagues could not accept women as equals and doubted women's capabilities as scientists, field researchers, engineers and teachers of male students.

---

[22] Fu and Wang (1996) cited in Wang and Stocker (2010).
[23] Cheng 2010. A similar experience has been noted among Indian women scientists who begin to publish in significant numbers when they are in their fifties after their children have passed the crucial high school age, which in India involves studying for challenging college entrance examinations that impose considerable stress on parents.
[24] Cheng 2010.
[25] Acar 1990. Gupta and Sharma (2002) report that all the Indian women scientists they interviewed were employing either full-time or part-time paid domestic help to enable them to maintain their busy teaching and research schedules.
[26] Gupta and Sharma 2003, 2002; Subrahmanyan 1998; Gupta 2016; Kumar 2001; Gupta 2007a and 2007b.

Indian women scientists report feeling isolated and excluded from informal male social groups at work; 70 percent have no female mentors. The idealization of the perfect scientist as one who works long hours and manifests complete devotion to "his" science further privileges Indian male scientists who can count on others in their household to take care of all domestic responsibilities for them. Indian women scientists also report turning down invitations to international conferences and fellowships because of their family commitments.[27] They tend to work in smaller teams, have fewer international collaborative papers, publish in relatively low-impact-factor journals, and are also not as well-cited compared to their male colleagues – all of which feed into a vicious downward spiral of less recognition and respect for these women scientists.[28] Still, there has been some progress in recent years, with roughly 40 percent of undergraduate enrolment in the sciences and mathematics in India in the 2000s being women.[29]

Singapore is in a significantly better position, with 60 percent of the students in the country who are pursuing full-time tertiary science degrees in 2015 being women.[30] Though only 27 percent of female science graduates in Singapore go on to pursue careers in STEM, the proportion of female research scientists and engineers in 2015 was 29.4 percent, a slight improvement since 2011 when the proportion was 27.6 percent.[31] Aside from these statistics, very little other research exists on the state of Singaporean women scientists, though there do exist several support groups in the country for women scientists.

Asian Women Scientists through the Life Course

In order to address the general lack of research on Asian women scientists' experiences, this chapter focuses on the various gender-specific factors that influenced the lives and careers of the forty women scientists

---

[27] Gupta and Sharma 2003, 2002; Krishna Raj 1991; Chakravarthy 1986.
[28] Garg and Kumar 2014.
[29] Gupta 2016.
[30] "Education and Training: University – Science, " Ministry of Social and Family Development, accessed on May 23, 2018, www.msf.gov.sg/research-and-data/Research-and-Statistics/Pages/Education-Training-University-Science.aspx.
[31] "Labour Force and the Economy: Research Scientists and Engineers," Ministry of Social and Family Development, accessed on May 23, 2018, www.msf.gov.sg/research-and-data/Research-and-Statistics/Pages/Labour-Force-and-the-Economy-Research-Scientists-Engineers.aspx.

Table 6.1 *Descriptive statistics of female interviewees, by country of interview*

Country of interview	Number of women (total)	Number of married women	% married to another scientist
China	2	2	100%
India	5	4	50%
Singapore	9	5	80%
Taiwan	9	7	71%
United States	15	13	69%
Total	40	31	71%

I interviewed. (Table 6.1 provides a breakdown of their origin countries and marital status.) I start with their lives as students of science, and later as practitioners of science, exploring their decision-making processes regarding where to seek training and employment. The idea of gender shock came to me during these interviews. While both men and women scientists recounted experiences of culture shock with each of their migrations, only women spoke of their gender as having a discernible impact on their professional lives. Men did not speak of their gender at all because their gender privilege was largely invisible to them, while the women I interviewed spoke frequently about how their gender started influencing their careers relatively early in their training, well before they left their home countries. While analyzing my interview transcripts, I paid specific attention to the ways in which gender did or did not play a role in interviewees' decision to seek training in the West, become an academic scientist, return to Asia or not, and finally, their decision about how high to aim in their careers.

Even though I was interviewing elite women scientists who studied and worked at some of the top science universities and institutes in their respective countries, they still described experiences of gender shock. It is likely that the accounts of women scientists from less elite (and possibly less egalitarian) institutions would have been even more extreme. But the fact that there was still a stark disparity between the accounts of the men and women I interviewed highlights how their training and employment in elite institutions and marriage to equally educated men, did not completely shield these Asian women scientists from gender shocks at school, work or home.

I found that in responding to these gender shocks, the women scientists I interviewed often reacted by making a gender compromise: settling for a less-than-preferred course of action in their career, marriage, or some other domain of their life in order to satisfy gender-related pressures in the same or another life domain. I coined this idea of a gender compromise by drawing on the idea of a "masculine compromise" put forward by Hong Kong-based sociologists Susanne Choi and Yinni Peng (2016). Studying a very different population, Choi and Peng describe the masculine compromises made by Chinese rural-to-urban male migrants in contemporary postsocialist China to preserve their symbolic status within their household in the face of rapid social change, economic precarity and Chinese women's increasing labor force participation. The authors highlight how the male migrants they interviewed made concessions on "marital power and domestic division of labor," while still seeking to maintain some kind of gender boundary between "big," "outside" and "heavy" tasks versus what they have redefined as "trivial" tasks suitable for women (Choi and Peng 2016:152). In Choi and Peng's example, these Chinese men carve out certain domains of their life where they are willing to compromise their gendered expectations or standards for the sake of family happiness or simple survival. I open up Choi and Peng's notion of masculine compromise to make it the more gender-neutral gender compromise, recognizing that, more often than not, it is still women who make concessions, in order to reconcile competing gendered pressures in their lives. However, what I also try to show in this chapter is that gender compromises on the career front are just one out of a range of concessions that Asian women scientists may make as they agentically navigate their intersecting worlds of work, education and family. In addition, I show that it is also possible for the *husbands* of Asian women scientists to make gender compromises at times. But before I get to the gender compromises made by my female and male interviewees, I start with a description of Asian women scientists' younger days when they first fell in love with science, and particularly, biology.

Daughters of Science

I went into this research project imagining that many of the Asian women scientists I interviewed would tell me that they had been discouraged from pursuing science by parents who insisted that science was a man's field. I expected these women to tell me that their status as daughters had meant that their parents had objected to their studying abroad. So I was surprised to never hear this said.

The Asian women scientists I met tended to come from relatively privileged sectors of their home societies, more consistently so than the male scientists I interviewed. These women were often the daughters of people with tertiary degrees and/or had grown up in urban areas where they were exposed to more educational opportunities. In this regard, they were similar to Nobel Prize-winning women scientists, all of whom came from "professional or academic families."[32] The parents of Vanya, a scientist from India, were both scientists who had lived in the USA for several years before she was born. Vanya grew up on a research university campus where she was exposed to conversations about science and scientific research from an early age, and was organically drawn to biology as a career. Both the parents of Sakuko, the Japanese scientist whose story started this chapter, held degrees in science and encouraged her to consider the USA when she first expressed a desire to also pursue a career in science. The parents of Rui, a Taiwanese scientist, were both scientists who received part of their graduate training in the USA before returning to Taiwan. Melissa, a Singaporean scientist, had been educated in a series of international schools, as her father worked in different countries throughout her childhood. The parents of Ting-wei, a Chinese scientist, were high school teachers – one in mathematics and the other in physics.

Only two of my female interviewees came from objectively poor families. Shirlena's father had worked as a bus driver in Taiwan, while her mother had worked as a street cleaner and domestic worker. But Shirlena still spoke of growing up in a family environment where education was prized as a noble pursuit, regardless of gender. Mishael, a Pakistani scientist, was also greatly supported by her single mother:

> She really supported and educated both me and my brother. She didn't have any real big ambitions for me, but she was very pro-education and she realized that women are much more vulnerable than men. ... She pushed me to have a career, but the choice of career was left to me.

[32] McGrayne 1998:6.

Paradoxically, some women told me that it was their "lesser" status within their family which led their parents to give them more latitude when it came to their career choices. In some families, daughters are understood to be "lost" to the family because they will marry and become part of their husband's family, and so there was less concern about what they did and where they went. Such an attitude had encouraged some of the women I interviewed to be more open to migration. Aiko, who trained as a medical doctor in Japan before going to the USA, told me that her status in her family oriented her towards thinking of leaving home from an early age:

> Firstly, I was a second daughter in the house, right? So I was nobody. ... If you are the first-born son, the parents tell you, "Stay home to take care of the parents." But if you are the second daughter [like me, they say], "When you grow up, please go somewhere." So I think, mentally, I was very prepared to go.

Like Aiko, other women who were younger siblings explained that they were given a longer window of time (and concomitantly, more freedom) during which they could pursue their dreams of graduate study in science. An Indian scientist, Aparna, now a professor at an Ivy League university, came from an all-girl family. She explained that her parents had been focused on marrying off her oldest sister, leaving Aparna free to pursue graduate school in the interim:

> You know how things work in India! It wasn't as though my parents would start thinking about arranging a marriage for me before [my sister]. But also, maybe it had to do with the fact that we don't have any brothers. My sisters and I were raised more to make a way for ourselves, get degrees that would help us do that, and not just go to college and get married.

In contrast to the national-level data I presented earlier in this chapter, which points to significant shortfalls in women's pursuit of STEM careers across many Asian countries, these early stories from my interviewees demonstrate that the contemporary Asian scientific field is not uniformly biased against aspiring women scientists. Asian women born with some privilege (as a result of their parents' socioeconomic status and educational attainment level) are able to take advantage of their relatively greater resources to pursue their dreams of undergraduate and graduate study in science. But what the following sections also show is that these women's relative privilege does not completely shield them from gender discrimination and the need to make gender compromises later in life.

Studying Science in Asia

While all my female interviewees spoke of supportive parents who encouraged their love of science and pursuit of a scientific career, interviewees' accounts of studying science at the undergraduate and graduate levels in their birth countries included some negative experiences. It was at this point in their education – when the idea of an academic career in science began to become more concrete in their minds – that several interviewees reported their first experience of negative gender shock and a recognition that their gender was negatively impacting their learning experience.

Shreya, an Indian scientist, spoke of having had wonderful biology teachers (both male and female) throughout primary, middle and high school, as well as throughout her undergraduate studies in India. It was these nurturing and inquisitive teachers who sparked in her a love of the subject:

> Biology was my favorite subject in school. I had a teacher who taught us biology in a most loving way. She reached out to the students, showing us, "This is a garden pea, green is its color." She was always telling us interesting stories. From kindergarten to high school, I had really good teachers in biology. And our school was really great in biology.

This supportive environment continued throughout Shreya's undergraduate studies. It was only when Shreya reached the doctoral level in India that she had her first encounter with gender discrimination. She had won a prestigious scholarship to pursue doctoral studies at one of the top science institutes in the country and was placed in a laboratory belonging to a male professor whose work she had long admired. Like Sakuko, she was the only female student in his lab. Very quickly, she ran into problems with her supervisor:

> My experience killed all desire to do research in India.... I think one aspect of it is the male ego. That triggered all the problems

that happened to me. They had this idea that "females shouldn't do science."

Over time, Shreya realized that it would be impossible for her to find a postdoctoral fellowship in India because her PhD advisor was giving her negative reviews whenever anyone called him for an assessment of Shreya. She was fortunate enough to be directly approached by a US professor who had come across her published work. He offered her a postdoctoral position in his lab, allowing her to leave India while remaining in her chosen subfield.

Like Shreya, several female scientist interviewees spoke of the challenges they experienced pursuing their interest in science at the graduate level in their Asian home country at a time when they had few female peers and almost no female science professors to interact with. Sakuko spoke of how, when she began her doctoral studies in Japan in the mid-1990s, she joined a lab where she was the first ever female student:

> My parents raised me, not really treating me as a boy or a girl. So I was totally unaware of my gender. I mean, of course, I know I was a female but I never thought that's going to really influence my life or my decisions. My undergrad was an open environment. It really doesn't matter, but ... my professor for my PhD had had his lab for over twenty years. And I was the first female graduate student in his lab. ... And then, almost immediately, I realized, "Oh, my gosh, this is a harsh environment."

Sakuko explained that there was initially only awkwardness from the male students in the lab who did not know how to treat her as a peer. From her professor, however, there was increasing denigration. "My passion toward science was totally going down and I thought I'm not made for science at all," Sakuko shared. Note that Sakuko experienced her first gender shock in Japan itself, having transitioned from a childhood and youth in Japan where she had not felt any handicap from being female. This speaks to the danger of painting all of Japanese society (or, for that matter, all of Asia) as sexist, when in fact gender bias is concentrated in particular social institutions and particular temporal moments in the career course of women scientists. Still, it was only after Sakuko moved to the USA (on the urging of her Japanese boyfriend) to pursue postdoctoral training in a lab run by an American woman scientist that she rekindled her love for science and rebuilt her self-confidence.

However, most of my female interviewees did not speak of negative graduate experiences in their home country that were directly related to their gender. Their status as *students of science* largely shielded them from such experiences because they were in a legibly lower rank within their university hierarchy with a clearly defined role as students, making it easier for their male peers and teachers to know how to interact with them. While even the idea of women studying science in Asia may have been considered transgressive in the past, sufficient progress had been made in my four Asian case countries such that most of my interviewees encountered only limited resistance when they were pursuing undergraduate and graduate studies in their home countries.

Heading West

As with their male compatriots, there came a time when the Asian women scientists I interviewed had to decide whether or not to follow their passion for science and seek training in the West. In most ways, these women scientists' accounts of their migration decision-making were similar to the accounts shared by the male scientists I interviewed. Like their male counterparts, aspiring Asian women scientists were drawn to the West because of the superior scientific infrastructure and training that was available to them there, the ability to boost their academic credentials, and a chance to experience what they understood to be a more cosmopolitan modernity than what was available to them at home. But one unique factor that drew many of my women interviewees to the West was the hope of experiencing a world where their gender would not serve as an impediment to their career progression, as several had already begun to intuit that they were going to face restricted training and career opportunities at home.[33]

While their gender encouraged their westward migration aspirations, it also *complicated* their migration decision-making process by

---

[33] Female brain drain has been shown to be significantly higher than male brain drain from developing countries, and it has been surmised that highly skilled women are more likely to desire leaving their birth countries if they do not see their human capital being adequately recognized and rewarded because of gender bias at home (Docquier, Lindsay, and Marfouk 2008).

drawing in their parents and partners into the deliberations, in ways that did not apply to my male interviewees. I am not implying that my women interviewees' social networks were not supportive, but rather that my female interviewees tended to give greater consideration to their parents and partners' opinions and preferences than my male interviewees ever did.

This greater consideration manifested itself in small and big ways. One way was in the funding of migrations. A greater proportion of my women interviewees received government scholarships to study in the West, and these scientists spoke with great pride about not having asked their parents to underwrite their dreams of a scientific career. None of them directly stated that their parents would have been unwilling to fund their overseas education, but it is telling how relieved these women were that they did not have to ask their parents for money. Inez received a prestigious government scholarship that allowed her to study overseas, after she failed to gain acceptance into Singapore's only medical school at the time – an institution where there had been a long-standing bias against women applicants, driven by a concern that women would drop out of medicine after they had children (and thereby "waste" the government's investment).[34] Inez shared that she had felt guilty about asking her parents to fund her overseas education, when they also had to take care of her younger brother's education needs. By winning a scholarship, she was able to fulfill the role of a "dutiful daughter"[35] who did not become a financial burden on her parents and instead aided the family at large. Rui, from Taiwan, received a full national scholarship to pursue a PhD in the USA, and shared that her parents were "proud that I could have a scholarship so they don't need to pay anything." Meanwhile, Fei Hong, a Chinese scientist now settled in the USA, had won a scholarship that allowed top biology students to pursue doctoral degrees at select American universities. Fei Hong's father had died when she was young, and she shared that she would never have left her widowed mother alone in China if it were not for winning this scholarship.

---

[34] Until 2002, a cap existed on the number of female students that could be accepted into the National University of Singapore's medical school, the only medical training program in the country at the time. The argument behind this cap was that women doctors' attrition rate was considerably higher than that of male doctors (Tambyah 2005).

[35] Oishi 2005; Paul 2015.

In a handful of cases from India, interviewees' gender *aided* their westward migration, as the institution of arranged marriage offered these women the opportunity to move to the West on a spousal visa and then apply for graduate programs from the West itself. One Indian scientist, Meena, wed an Indian-American citizen who had been raised in the USA. Meena had already finished her master's degree in biology in India, but with her marriage and impending move to the United States, she decided to pursue doctoral studies in that country. While she might have done that even without marrying an American, her arranged marriage made it easier for her given that her spouse was already based in the USA. In addition, the fact that he was working and earning an income meant that she would not have to struggle to make ends meet as an international student. Aditi, another Indian bioscientist, had an arranged marriage with Shrikar, who was pursuing his PhD in a European country. Aditi was finishing a master's degree in biology in India at the time of her wedding and afterwards moved to Europe to join her new husband. Once there, she joined a PhD program in the same country as her husband.

This particular mechanism for entry into the West was only seen among my Indian female interviewees and it certainly facilitated their westward migration for further training. The disadvantage that came with this migration pathway was that it constrained these women's overseas location choices. Aditi might have considered PhD programs in the USA but, because her husband was in Europe, she limited herself to graduate programs in the same European country as him. Likewise, Meena only applied to doctoral programs in the US state where her husband was working. In this manner, family concerns were woven into my female interviewees' westward migration decisions. During my interviews, I heard multiple accounts of gender compromises being made by women scientists in terms of the location of their graduate education and the timing of their departure to accommodate the preferences of their boyfriend or husband who was also moving West or was already there.

One such story was told to me by Rohan, an Indian molecular biologist, who met and married his wife when they were both doctoral students in India. He shared how *his* location preferences had prevailed when he and his wife were considering where to move to in the USA for postdoctoral training. He was upfront about the fact that their eventual destination in the USA had not involved any compromise on his part, but had catered only to his preferences and not his wife's:

> We struggled a lot to find a place, to find the same city in the US because we didn't want to go to different cities for our postdocs. And, as it turns out, [my wife] ended up paying a big price for that. She got a lot of good [postdoc] offers because her PhD was very good. And, of course, she is a smart scientist. So she had good offers, but I was unable to find anything good. I had three, four good offers [from universities] where she was not able to find any offer. And then, as it happened, I got a fantastic offer and I say that based on the fact that I met the person and his list of postdocs who had moved out of his lab had gone to excellent places. He was not only known as a good scientist but also a good mentor. So it was like a dream lab. But then [my wife] didn't have any of her first- or even second-list of PIs there [at that university]. She had one offer from there which was not on her [preference] list, but I ended up convincing her that it won't be that bad. [That] she would do well there. But it didn't turn out that way.

Rohan's wife, Meera, ended up going through three postdoctoral supervisors at the university Rohan selected before she was finally able to find a professor with whom she could build a good working relationship. As a result, she experienced a significant delay in starting her postdoctoral research and subsequently experienced further hurdles in her career. Likewise, Aditi, who followed her husband to Europe, had her career continuously straitjacketed by his location decisions for his postdoctoral training and his first assistant professorship. I will discuss these negative impacts on Aditi's career later in this chapter. But first, I want to share what my female interviewees recounted about their years training and living in the West.

Rethinking Gender in the Lab

For all of my female interviewees, as it was for my male interviewees, moving to the West opened both their minds and their ambitions. In Chapter 4, I wrote about the new scientific cultures that interviewees were exposed to in the West. These scientific cultures were just as revelatory for my female interviewees. But once again, for my female interviewees, there was an additional gendered element to this experience. In several cases, they were able to work under senior *women* scientists, something none of them had experienced when they were in their home country. And because my interviewees had trained at top research universities in the West, the women professors they encountered were a highly selective group – ambitious, productive and operating at the top of their game. These female role models presented a vision to my interviewees of what it meant to be a successful woman scientist.

Even though none of these female scientist-mentors were of Asian descent, there was still a bond that developed between them and my female interviewees. Sakuko, the Japanese scientist who at one point had considered dropping out of science entirely, joined a lab run by a female "superstar" scientist at a top research university. This was Sakuko's first position upon her arrival in the USA, and she described the experience of starting work in that lab as life-changing:

> As soon as I started in that lab, oh my gosh, I really felt like my cells are breathing. [I thought,] "Oh my God, this is fun!" I was greatly accepted. So talking about the migration from the place you were born to somewhere else, there's always a balance. Like how much you feel comfortable in your home country versus where you move. How much distance can you tolerate being in a foreign place versus the benefit of being in a foreign place? But in my case, I was way more comfortable *not* being in my home country. Of course, I miss my family but, other than that, just the working environment, the research environment, I felt completely at home. [Earlier] I didn't know if I can really stay or survive in this country. But I knew within a week like, "Oh my gosh, I feel so comfortable here!" So that's when I thought, okay, I should probably look for a job in this country.

This was Sakuko's second experience of gender shock but, this time, it was a positive one that made her feel that her gender was no longer an impediment to her dream of becoming a scientist. Sakuko spoke in glowing terms of her postdoctoral supervisor, with whom she worked for five years: "I was very lucky because … my postdoc advisor turned out to be an extremely supportive person. As supportive as you can imagine. You know, she's an extraordinary mentor so I loved being in her lab." Likewise, Alisha compared the two mentors (one male and one female) that she had worked with in the USA during her doctoral training, and reflected on how her female mentor was the more ambitious one:

> They're both equally successful and famous but, in fact, my current [female] boss is even more famous and prolific and hardworking. ... I know there are studies out there that show that female scientists don't do equally well or progress equally fast because of family, child bearing, child-rearing decisions, and I guess that's true. And there's also studies that show that PhD students, if they're female, they will publish less avidly than their male colleagues, and most of it is attributed to being distracted by trying to have a family at the same time. So that might be true, but I don't see it in my experience [with my mentor] at least.

As a direct result of having this positive female role model, Alisha felt empowered to aim high in her own career too.

Interviewees like Alisha and Sakuko belonged to a generation of Asian women scientists who had hardly any female mentors in their universities in Asia, and so they were able to interact with strong women role models in science only *after* they moved to the West. They had all grown up reading about the Nobel Prize-winning scientist Marie Curie, but it was only when they moved to the West that they encountered high-achieving women scientists in the flesh, and were able to take classes with them or work beside them in the lab. The situation is different now in my four Asian country case studies, where there is a small but growing number of women academics running successful labs (including the women I interviewed who had returned to Asia). These women now serve as mentors for the next generation of Asian women scientists. As a result, in the present day, Asian female science students do not have to leave Asia to find female mentors and role models, or to work in labs where they are treated fairly and inclusively. As a result, the transition from student of science to practicing scientist is smoother for women in Asia in the twenty-first century than it had been in the past.

Marked by Gender even in the West

Despite having largely relished their time as students and trainees in the West, female interviewees who chose to embark on careers as independent scientists in Western countries quickly realized that gender discrimination and marginalization of women scientists happened in the West too. Aiko, an unmarried Japanese scientist, was frank about what she saw as the daily challenges faced by nonnative women scientists of color in the USA:

> In this country, people say you have to be three times smarter than Americans, you have to work five times harder than Americans. ... In the US, it is hard and I have to show that I'm good every day. [But] if you are white man, you just need to show only once, right? Instead, I have to show it every day and I have to fight every step of the way and then I still get the short end [of the stick].

Aiko emphasized how her intersecting racial and gender identities influenced how she was treated by her peers, but most female interviewees stressed that their gender had the greater impact. Like Aiko, Pakistani scientist Mishael was well aware of the gender bias that was rampant in her chosen subfield in the USA. She was not under any illusion about what her "greatest limitation," as she called it, was:

> Out of being nonwhite, not speaking with an American accent, being Muslim, being Pakistani, having an obvious Muslim name, and being a woman – out of all of those limitations, so to speak – the hardest is being a woman. Isn't this amazing? It's okay, if you're not white. ... Out of all those quote-unquote "limitations," the greatest limitation to deal with is being a woman. In Pakistan, the chauvinism is really in your face and obvious and irritating. But there are chauvinistic attitudes over here too which are much more subtle, but they are there.

To make matters worse, women scientists' transition to becoming independent researchers often coincided with a shift in their family status, since many also started a family at this point in their lives. In terms of how this change impacted their careers, most interviewees deemphasized the biological impact of giving birth and focused more on the social pressures to be the primary caregiver within their family unit. This societal expectation hampered several interviewees' career progression, even though such gender mores were less restrictive in the West as compared to Asia. Aditi, mentioned earlier, who joined her husband in Europe, was accepted into a PhD program in the same European city as her husband and had two children while she was pursuing her degree. Her husband finished his PhD earlier than she did and then moved to pursue a postdoctoral fellowship in another European country. Aditi did not follow him because she had to stay close to her doctoral lab to run her experiments. This meant that she was effectively raising their children by herself. Her husband would visit every two to three weeks, but his frequent absences took a heavy toll on their family. Their oldest daughter interpreted the transnational split-household arrangement as a divorce and became depressed. Eventually, the situation became so emotionally taxing for the family that Aditi temporarily gave up her doctoral studies to join her husband – a decision, she later admitted, that derailed her academic career:

> I have no regrets [about the move] because of the family part of my life. But professionally, yes, that was the worst decision of my professional life.

Aditi eventually found a new professor who agreed to supervise the completion of her PhD, but it took her another three years to finish, leaving her with so many gaps in her CV that it was hard for her to find a tenure-track position when her husband decided to move the family back to Asia. She eventually dropped out of science to become a full-time mother.[36]

Asian women who did not want to compromise their career ambitions for their husbands, found the traditional gender norms pervasive throughout many Asian countries to be suffocating. Among the male and female Asian scientists I interviewed who had chosen not to return to Asia, a higher proportion of the women were divorced or unmarried. Of the eighteen male scientists interviewed in the United States, one had never been married while all the others were married. In comparison, among the fourteen female scientists I interviewed in the USA, two had been divorced (though one of them had subsequently remarried) and a third scientist had never married. Alisha, a clinician scientist, bluntly explained the reason behind her divorce as professional jealousy on the part of her husband: "It was just [him] knowing that your spouse has one more thing that you don't," Alisha explained, referring to the fact that she had a PhD on top of her MD qualifications while he "only" had an MD. Postdivorce, she reflected that she had not fit the mold of the

---

[36] Aditi's story is emblematic of the "cracks" (Nowotny 2017) that can appear in women scientists' timelines when their life course and career course come into conflict with each other. In her case, the cracks in her career course were too many and too wide to be crossed and she dropped out of science.

stereotypical Asian wife who put her husband's needs and desires ahead of her own. Though he had initially supported her when she decided to pursue a PhD, over time his support waned. "Maybe I didn't pay as much attention," Alisha mused towards the end of her interview. "I got distracted by having too much on my plate and didn't give him the attention he needed." She went on to recount how her (Asian and non-Asian) male colleagues were often puzzled by her ambition: "Actually, one of the [surgical] fellows asked me, 'Why are you doing your PhD? You know you can only be called a doctor once?'" Such comments highlight the lingering discomfort with ambitious women scientists, even in the West.

Ting-wei, an assistant professor at an East Coast university, divorced her Chinese husband when they were both pursuing a PhD in the USA. Looking back, she concluded that she had married too young when neither she nor her husband knew what they wanted in life or in a partner. As her ambitions grew, she also grew apart from her husband: "I think he needs someone who's more willing to defer to him and I need someone who can support me and intellectually nurture me. There's just different needs." Both Alisha and Ting-wei had decided that they could not compromise on their scientific careers, and instead chose to jettison their marriages.

Unlike the men, the women scientists I interviewed spoke of having to choose which one of their "clocks"[37] to prioritize because they were well aware that they could not give adequate attention to all three – the biological, the research and the domestic. "Leaning in"[38] on all domains of their lives was not an option. Priyanka, an Indian scientist, decided to divorce her husband because she was unwilling to split her time between different priorities given that her first love was in fact science: "I didn't want to be held back by anything! Not family, not marriage – nothing." Priyanka saw her professional ambitions as incompatible with the social responsibilities of being a wife and mother, especially in India. For Priyanka, the gender compromise she made was to give up the labels of wife and mother – effectively to throw out two of her clocks – so as to focus exclusively on her life as a scientist.

---

[37] Stolte-Heiskanen 1991.
[38] Sandberg and Scovell 2013.

In this manner, as they moved through their intersecting life course and career course, Asian women scientists often found themselves making gender compromises. But what I show here is that their gender compromises were not always on the career front; instead, some women scientists chose to compromise on the family front. But whatever they chose, all my female interviewees understood that their gender identity meant that they were more likely to be the ones making compromises in one life domain or another.

The Question of Return

When the question of return came up for Asian women scientists, the impact of their gender became even more palpable as the prospect of return forced a direct comparison between the situation they faced in the West with what they expected to experience if they went back to Asia. They were in effect imagining what new kinds of gender shock they would experience and what new gender compromises they would have to make if they returned. It tended to be the case that the longer they spent in the West, the less interested my female interviewees were in returning. Having experienced working life in a more gender-progressive environment overseas (even with its ongoing gender biases), the Asian women scientists I interviewed were largely reluctant to return to what they imagined would be a more conservative space. They spoke of how they had changed as individuals during their years spent overseas and how their priorities no longer aligned with what they imagined they would encounter in Asia. Shu Jing, a tenured scientist at a US government research institute, spoke of how she was no longer a good fit with the scientific culture in China. Shu Jing reflected that she had changed during her many years in the United States, remarking, "I don't know if it is the USA or my profession, [but] I think I am changed a lot. I am more goal-oriented, more bold and straight to the point and open and, I think, more confident." For all these reasons, Shu Jing did not believe she would adapt well to a research environment where women were expected to be quieter, more submissive and less ambitious than their male colleagues.

Among the female Japanese scientists I interviewed in the USA, there was no question of returning to Japan. While they often went back to visit family or give talks and attend conferences, these women were adamant that the only reason they had risen in their careers was because they worked in a society that better recognized and rewarded their talents. Aiko told me: "Nowadays, I'm invited from Japan to give a talk or to do some collaborations, right? When I go back to Japan and then talk at a conference, all the other speakers are male. I don't see any females!" Aiko was cleareyed about the prejudices that existed within American academia but, in her mind, the gender and racial biases within the American scientific establishment were nowhere near as extreme as in Japan:

> I still have a chance to be a professor in this country[, the USA]. In Japan, there is no chance! Zero chance! I'm not even considered. And every time I go to the conference in Japan and I see only white and only Japanese men on the podium, I think I made the right choice [to stay in the USA].

Mishael also recognized that her time in the USA had changed her to the point where she would experience a negative gender shock if she returned to Pakistan:

> Growing up, there was a lot of chauvinism that you have to deal with, and I think I'm much less equipped to handle those kind of attitudes now than I was before. I think with age and an education, and maybe independence, I've just gotten more verbal and less patient with those kinds of attitudes. So I think I would get a culture shock if I go back. And vice versa, I would be extremely shocking to a lot of people [there]! I wouldn't be what they expect me to be.

Mishael was not interested in returning to Pakistan, but she was not married and so she could make this decision on her own. For *married* women scientists, the return decision was typically not one they were able to make on their own, as they were heavily influenced by their husband's preferences. So it helped when both husband and wife were on the same page about their location preferences and career goals. For Meena, who was married to an ethnic Indian man born and raised in the USA, there was no question of returning to India because her husband, who had never lived there, had no interest in living in her home country. Likewise, Ting-ting did not entertain the possibility of returning to Taiwan because of what it would mean for her husband, an ethnic Taiwanese medical doctor who had been raised in the USA:

> Did we ever consider moving back to Taiwan? No! My husband would not survive! I know that. He barely speaks any Chinese. He is an MD and so there is a licensing problem as well. I would imagine that that is very complicated. And he hasn't done his military service so I don't know how that would be handled.

In Meena and Ting-ting's cases, both husband and wife wanted to stay in the United States and so there was no conflict in their location preferences. When husband and wife *did* have opposing priorities, however, with the husband wanting to return and the wife wanting to stay put, the women I interviewed responded in differing ways. More senior women scientists who had received tenure or were more advanced in their careers tended to hold their ground and give their husbands an ultimatum. They either flatly said no, or they told their husbands to return to Asia if they wanted, while they (the women) would remain in the West with their children. This is what Hui Ting, who was a tenured research scientist in the USA, told her husband when he raised the possibility of returning to China to take up a full professorship. Earlier in their relationship, Hui Ting had allowed her husband to dictate their family moves – first from China to Europe for graduate studies, and then from Europe to the USA. She had followed him because he was a couple of years ahead of her in terms of his training and she was initially unsure as to what she wanted to do with her career. But once she settled into her research career and received tenure at an American research university, she had no interest in giving it up. When her husband received a tenured offer from a top research university in China, he considered accepting it, but Hui Ting chose to put her foot down:

> We didn't argue but I did tell him my attitude. I told him that if he decided to go back, we better divorce first. Then he can pursue his future with anybody else. I'll stay here and continue my life here [in the USA]. But he didn't like the idea. So that's the condition. He can go back but we have to officially divorce first. Because many people go back and they have a second family there, you know? That was common and I didn't want that to happen. So then he decided to stay.

The fact that Hui Ting had already received tenure in the USA boosted both her self-confidence and also her willingness to insist on her personal location preference in a way she had never done earlier in her marriage. My interviews in the USA and Asia revealed that women often tended to give in to their husband's desire to return home when they were still new to married life or were at an earlier stage in their career course compared with their husband. Earlier in her marriage, Hui Ting had left a tenure-track position and followed her husband to a different US city where he had received a tenure-track offer, just so that the family could stay together:

> Was I upset about moving? Yes and no. I'm still a very traditional Asian woman and so I wasn't *not* going to move.... And we have three children so we didn't have any debate about whether or not I should move.

Contrast Hui Ting's earlier position with how she later insisted on the primacy of her career over her role as a wife. As a result, she was able to force her husband to make a gender compromise in his career course to keep the family unit intact. It was clear to me from such accounts that Asian male scientists do make gender compromises as well, but that it often requires their wives to reach a significantly advanced stage within their own career course before the scales tip in the woman's favor during their family negotiations. A male Chinese scientist, Seng Kiang, whom I interviewed after he had returned to China, shared that his ex-wife (also a bioscientist) had refused to return to China with him when he was offered a professorship back home, because she had already been awarded tenure at a midwestern research university in the USA and did not relish the idea of moving her research back to China. They subsequently divorced, and their son stayed in the USA with his mother. In fact, several of the senior-level, returning male scientists I interviewed in Asia had left their wives and children in the West and were living in a split-household arrangement because their wives (all of whom had successful careers of their own) had chosen not to move back to Asia with them, or had tried it for a while and then decided to return to the West on their own.

Women scientists who were more junior, however, often returned to Asia to support their husband's career priorities, even to the detriment of their own careers. Rohan, whom I interviewed just a few months after his return to India, spoke of how all of his and his wife's moves, from the time they first left India to pursue postdoctoral positions in the USA, had been for his benefit. I have already discussed how Rohan's wife, Meera, faced difficulties finding a suitable postdoctoral supervisor when they first moved to the USA. Eventually, however, she was able to find a supervisor with whom she worked well. But before she could successfully complete her postdoctoral training, Rohan was offered a tenure-track position back in India. They returned to India without Meera having the necessary publications that would secure her a good academic position too, and she thus found herself in academic limbo back in India. Rohan was very aware that his preferences had determined all their moves: "From the time we left [our Indian PhD university] and went to the USA, it's like we had two contrasting lives," he shared with me, guilt threading his voice. "You know, everything went very well for me. The postdoc and subsequently getting a job, and [meanwhile] everything almost didn't go well in her case, except that we have two beautiful kids." Although both partners had started their scientific careers at similar points, Rohan's career and migration preferences had repeatedly taken precedence over his wife's, leading to a sharp divergence in their career prospects.

Gender Compromises upon Reentry

Returning women scientists typically found themselves joining Asian research organizations that had little female representation in their leadership ranks. Vanya spoke of how there was an ongoing dearth of women in the upper echelons of her Indian research organization, which made her hyperaware of her gender at work:

> You're really aware that you're a woman. That's for sure. Because the number of women in an institution like mine is only a handful. You find yourself in a committee and you're wondering, "Am I the token woman being put on this committee because you needed a token woman? Or are you putting me on the committee because you actually need my input?" So that is the reality.

Vanya noted that she had only been able to do well back in India because she had tremendous support from her husband and extended family so that she could devote as much time as she needed to her research.

Women scientists like Vanya who returned to Asia spoke of having to readjust, not just to the "new" scientific cultures in their return destinations but also to the way gender roles and norms played out in the broader society of their return destination. In Chapter 8, I speak to the new scientific cultures all returnees (male and female) encountered in Asia. In this present chapter, however, I focus on the particular gender shocks returning women scientists experienced and the gender compromises they were forced to make. Here there were differences in the experiences recounted by women scientists in India, China, Singapore and Taiwan. While all returned women scientists complained about organizational discrimination and marginalization, most of the complaints about gendered family norms came from East Asia. Broad-based societal discrimination was only mentioned in India.

Divya, who had completed her PhD and a postdoctoral fellowship in the USA before returning to India, spoke of how it was not the different scientific system in India that was the hardest to get used to but rather how gender operated socially and culturally, because that was where she saw the greatest divergence from her life in the USA:

> As a woman, you get used to certain freedoms in the Western world which can [make it] kind of hard coming back. ... One learns not to do certain things. Like walking on the street. It's just a bit of a challenge walking on the streets of India. And you have to get used to the idea, for instance, of not wearing certain kinds of clothes that I would be very comfortable wearing in the US or being out late by myself. Like the idea of traveling by myself now suddenly has to change. So I can't go exactly where I please, when I please.

I found myself surprised at how resigned Divya was to the idea of giving up the simple act of walking on the street or wearing certain clothes now that she was back in India. But then again, as an Indian woman myself, I fully understood what she meant, given the high rates of harassment and what is known in India as "eve teasing" that occur in certain parts of the country.[39] This is a reality that all Indian women learn to live with and navigate. But Divya also pointed out that because she spent so much of her daily hours in her lab, where she was in complete charge, she did not have too many other adjustments to make after her return to India.

Divya's account highlights both the ongoing gender prejudice women in India experience, as well as the relative privilege she enjoyed because of the rarified environment in her particular workplace. The

---

[39] Natarajan 2016; Ghosh 2011.

research organization where she worked was not emblematic of most Indian universities and the work environment was more progressive (at least on a day-to-day basis) than the average Indian university. Research by Namrata Gupta and other gender scholars at various tertiary institutions and across various scientific subfields in India reveals however that gender still plays a critical role within the Indian workplace.[40] Several of my female interviewees in India admitted that they had only applied to a select handful of universities and institutes – those considered the crown jewels of India's scientific establishment – and would not have considered returning at all if they had not been offered positions in one of these institutions.

In addition, Divya and her husband did not have children. Returning women scientists who were also mothers experienced another kind of gender shock when they found themselves solely responsible for many more domestic duties than they had had to take care of in the West.[41] The social roles of wife and mother were defined differently in Asian countries and were harder to juggle alongside the demands of academic life, even when a woman scientist's husband was also an academic. This was the biggest area of dissatisfaction that returning women scientists who were also mothers highlighted to me: how their husbands often fell back into more traditional social roles within the household and expected them to pick up the slack. The social and familial pressures to manage all domestic responsibilities appeared to be stronger for Asian women scientists in Asia, as compared to Asian women scientists in the West.[42] Runchen, who completed her PhD in Europe and then returned to Taiwan where she is now a senior scientist at a top research institute, spoke of the relatively low numbers of female PIs at her institute – a pattern that repeated itself throughout most of the Asian research organizations where I conducted interviews. She noted that "in university [in Taiwan], most of the female students got a good score, a better score even than the men," but later in their careers, men tend to overtake women. Runchen posited that this partly had to do with how Taiwanese women had more family commitments,

---

[40] Gupta 2020.
[41] See also Bhatt 2018 for a similar account from Indian women married to IT executives who returned to India from the USA.
[42] As in Chapter 4, this insight speaks to the malleability of individual behaviors and norms, and how the broader society and culture can create pressures on individuals to behave in particular ways.

which kept them closer to home and prevented them from prioritizing their career. She interpreted her own career through this lens:

> Even now, I'm still not very active [in my career]. For example, I won't go outside [the country] to give lectures or attend a meeting because I have kids to take care of. Most of the female scientists I know, they also are bounded by family, so they rarely go outside. And somebody, even if they don't have kids earlier, maybe when they graduate, they're probably doing their application and probably thinking whether they should have kids. And they delay the whole process. And when you do have kids, you might just think, "Oh wow! Maybe I should stay at home." Men, they don't have to worry about that, but we do.

Runchen admitted that her husband's lack of support for her career had dampened her ambition to the point where she had lost her drive to rise up the career ladder at her research institute. She did not see the point in being more ambitious since her husband did not appreciate having to take care of their children while she was away for work:

> My husband won't be happy if I go outside [the country] to attend meetings alone and leave him to take care of the kids. He loves his kids very much, but he likes to sit on his couch and just say, "[Runchen], you should do this, do that." And when a kid is crying, he just calls my name. He just sits there, watching his iPad! But I have to be there. I always think I'm just like the nanny.

For Runchen and most of the Asian women scientists I interviewed, marriage and family life represented a pull away from their research, and this pressure was ratcheted up if they returned to Asia, where there existed broader societal pressure in all four of my case countries to live up to an ideal of wifehood and motherhood that left little room for an ambitious scientific career. This gendered pressure experienced by women scientists was compounded by the more gender-neutral challenges of working in various Asian countries. I pick up the broader discussion of the various challenges faced by returning Asian scientists in the next two chapters.

Still, there were also personal and professional rewards that came from being back in Asia. Many of the returning women scientists I interviewed noted how they had a high proportion of female students and staff in their labs. Vanya, in India, shared that she had four postdoctoral trainees working with her in her lab and that all four were women. Other returned women scientists shared that they intentionally recruited more female students and trainees to work in their labs, partly because they appreciated being able to mentor these younger women, but also because it ensured that there were fewer tensions in their lab as some male students or trainees might not appreciate working for a female PI. Interviewees were clearly "paying it forward," serving as mentors to a new generation of aspiring Asian women scientists and demonstrating how it was possible to have a productive scientific career as an Asian woman in Asia itself.

Conclusion

Wherever they ended up – remaining in the West or returning to Asia – the women scientists I interviewed understood that their gender often prevented them from living up to their full potential as scientists. Aparna, an Indian molecular biologist in the USA, explained it in this manner:

> Being a woman has definitely impacted [my career]. Me being an Indian woman too, but I really think that it's the woman part that makes a difference more than the Indian part of it. It's such a male-dominated world and a lot of times I feel like I'm not being taken seriously. I think a lot of that has to do with [my gender]. I mean a lot of it is very subconscious, but it is there, I think. The more you go higher up, the more you encounter this.

For those who returned to Asia, where their ethnicity was less of an issue, it was their gender that again stood out as having a discernible impact on how they were treated in the workplace, in society and within their families. Still, the accounts I have shared here in this chapter point to how the present-day terrain for aspiring women scientists in various Asian countries is improving. For Asian women who are already advantaged with educated or middle-class parents, the path to scientific training is now smoother, and their gender represents less of a barrier than it used to. However, while family influence at the student stage in their careers can be largely supportive, especially when buffered by class status, there remain two inflection points in their intersecting life and career courses when women's gender becomes a significant obstacle to their success in the global scientific field.

One shift occurs when embarking upon an independent scientific career in a university or research institute setting. Asian women who had been accepted as students of science found that these male-dominated spaces were not as welcoming as they could be to female practitioners of science. This phenomenon is not unique to Asian women scientists or to Asian scientific organizations. However, research organizations in the West have been grappling with the question of gender mainstreaming for much longer than their peer institutions in Asia and, as a result, they have already put in place hiring and workplace policies that attempt to be more supportive of women scientists. Until the work environment in Asian universities and research institutes catches up, top female students from Asia will likely continue seeking training opportunities in the West in the hope of encountering a more gender-neutral environment.

Once in the West, Asian women may also deem their chances of academic and professional success to be stronger in Western institutions as compared with Asian ones, thus dissuading many of them from returning to Asia. However, as more Asian women scientists do return to Asia, either as independent migrants or as trailing spouses, there will be increased pressure on Asian research organizations. Once these countries reach a critical mass of women scientists, this will result in such conversations becoming louder and more impactful. Through my interviews with returning women scientists, I could already see the difference they were making as they served as mentors to new cohorts of aspiring Asian women scientists *in Asia itself*. In addition, these returning women scientists were not only positively influencing female students. They were also demonstrating to their male students and peers just how capable and high-achieving women could be.

The second inflection point Asian women scientists experience is when they shift to married life, which also introduces new gendered pressures into their lives. The social roles of wife (and later, mother) are hard to juggle with the demands of academic life, even when women's husbands may also be academics. Living and working in the West, where more liberal views often prevail about women's ability to pursue both family and career ambitions, may ease some of this struggle. The fear of a negative gender shock on the family or household front if they returned to Asia can keep ambitious Asian women scientists from wanting to go back.

Throughout their careers and migrations, the Asian women scientists I interviewed were making gender compromises more frequently than their male peers. These included prioritizing their partner's location preferences over their own, instituting a split-household arrangement, dissolving their marriage, or sidelining their research career as they attended to their domestic responsibilities. Some of these gender compromises prioritized their careers over family, while other compromises did the reverse. But all of these compromises were made with the recognition that the domains of family and career were intimately intertwined, and that it was not possible to maximize outcomes in all life domains simultaneously.

There were some women who were willing to scale back their research ambitions if that meant that they would have more time with their children, or because they prioritized their husband's career goals over their own. Over time, compromises involving the prioritization of family responsibilities and their husband's career and migration preferences, led to a cumulative disadvantage[43] for women scientists within the domain of their research careers. My interviews reveal that this cumulative disadvantage was often an iterative and layered process involving small gender compromises that over time added up to have a substantive impact on interviewees' scientific careers. Other Asian women scientists made different gender compromises, choosing to prioritize their careers or shield that domain of their lives from encroachment by competing marital or household demands. At the most extreme end were women who chose to dissolve their marriages, not marry at all, or not have children, so that they could devote more time and energy to their careers. Less "radical" solutions included refusing to leave the West after their training was completed and choosing to work there, rather than return to Asia. All of these decisions are gender compromises that reveal the interconnectedness of the different domains of Asian women scientists' lives.

The Asian women scientists I interviewed were fully aware that they were not able to "do it all" and that in order to "lean into" their careers, they would have to step back from other domains.[44] But

---

[43] Purkayastha 2005.
[44] Slaughter 2012; Sandberg and Scovell 2013.

because of their well-paid careers, many of my interviewees were able to access a wider range of gender compromise options – including remaining single, initiating a divorce, transnational householding, outsourcing their household responsibilities to a domestic worker, and refusing to move countries – than women with more precarious employment or less elite status.

At one level, these findings help explain the ongoing "leak" of women scientists from productive careers in science. But this issue is only partly a "science" problem. By considering *both* the career course and the life course together, this chapter highlights how the domain of the family directly impacts Asian women scientists' career and migration decisions, and vice versa. This chapter also highlights the social changes currently underway throughout Asia as Asian women enjoy an increase in their educational attainment levels and growing entry into the formal labor market. The fact that more senior women scientists were able to insist on their own preferences when negotiating with their husbands about possible family moves speaks to how women can accumulate symbolic capital within their household over time thereby enabling themselves to have more say in family decisions.

Finally, this chapter demonstrates how migration and destination decisions are intricately featured in Asian women scientists' gender compromises. With the globalization of science, there are increasing mobility expectations in terms of education, training and career progression. Married junior women scientists in particular may find themselves either involuntarily mobile or immobile to the detriment of their careers. Or they may decide that they cannot fully realize their ambitions back in Asia and so decide to leave and not return. Without concerted efforts to negotiate a new, more equitable gender contract within their families and Asian research organizations, Asian women scientists will continue to be underrepresented in the upper echelons of science in Asia.

PART III
Consequences

7 New Scientific Research Systems in a Changing Asia

> In the US, doing science was pretty easy. There was access to [supplies], the library, people who would help me, and so on. But there was no excitement in doing fieldwork.
>
> In India, whenever I went to any forest – even if a degraded forest, even if a fragmented forest – it was heartbreaking to see ... all the environmental and ecological destruction. It was heartbreaking, but at least I felt something about the things around me. There was excitement when I saw something new. There was immense tranquility when I just sat on a hillside in the evening. There was happiness whenever I went there.
>
> In the US, maybe sometimes I felt excited because I was seeing new things, but otherwise there was no excitement in anything I did. Life was comfortable, but it wasn't exactly happiness. It wasn't contentment. Whereas in India, even if things were hard, I always felt like I was living a life I liked. [In the USA,] it was comfortable, which a lot of people equate to being a good life, but it wasn't a good life for me.

The account above comes from Karthik, an Indian ecologist who completed his doctoral and postdoctoral studies in the USA before returning to India. Karthik, like all my other returnees, had occasionally visited his home country while he lived in the West. These visits were not just to see family and friends but also to attend conferences, start research collaborations or conduct fieldwork. These trips had given him a sense of how the research system in India had evolved since he first left for the USA.

Still, short-term visits are different from moving lock, stock and barrel to Asia, with one's family in tow, to set up a new lab in a new place. As described in his quote above, Karthik knew that returning to India would be challenging after the comforts of research life in the USA. But he decided that the rewards of being based back in India would make it worthwhile. So this chapter picks up where Chapter 5 ended, reconnecting with Asian scientists like Karthik who returned to Asia. In this chapter, I describe these scientists' *postreturn* experiences of "doing science" back in Asia and how they engaged with the scientific research systems in their new places of work.

These scientists' return was made more momentous because the "place" where they were returning had changed radically while they were away. This chapter documents the changes that returnees observed in their respective Asian country's research systems and how these changes affected their experience of conducting scientific research in Asia. While Chapter 2 described the recent scientific developments that have occurred in my four Asian case countries, it offered only an overview of the revised budget allocations, university restructurings and national R&D plans implemented by each government. This chapter takes a different approach, describing the on-the-ground, lived experiences of these changes as recounted by the returned Asian scientists I interviewed. Because these scientists had been overseas for ten years on average and were now embedded within the most elite organizations of their respective Asian country's scientific establishment, they could provide a firsthand account of how much the research systems in the country had changed since their initial westward migrations. Just under 20 percent of my returning interviewees had returned to Asia in the 1990s or earlier, before any large-scale changes began, and had personally helped to spearhead some of these developments. Close to 45 percent of returnees had returned in the 2000s when changes were being introduced in all four case countries (see Table 5.1). And another 38 percent returned in the 2010s when changes were picking up pace. As a result, the accounts from these three cohorts of returnees offer a sense of how the scientific terrain in select parts of Asia has been rapidly changing during the first two decades of the twenty-first century. Returnees' experiences in both Western and Asian countries also gave them an immediate comparative perspective that allowed them to explore what had improved, and also what had not improved, in Asian research systems since the turn of the century.

Returnees also spoke to the impact of these changes on their research productivity, though their experiences were limited to the top institutions in their respective country and so they could not speak to conditions in lower-tier research organizations in the same country. This selection bias is important to keep in mind when considering the areas where returnees noted that significant improvements had occurred, but also areas where they said that not enough change had happened.

On most dimensions, returnees found the research conditions in their return destination significantly improved, thereby positively influencing their postreturn research experiences. I call these research conditions at the return destination the "scientific contexts of return." Here, I am taking a page from the extensive literature on immigrant assimilation that attempts to theorize the conditions under which a first- or second-generation immigrant achieves socioeconomic mobility and sociocultural integration in their host country.[1] Segmented assimilation theory highlights how varying contexts of reception in the host country (receptive/indifferent/hostile government policy, prejudiced/nonprejudiced societal reception and a weak/strong coethnic community) interact with immigrant characteristics to create divergent pathways for different groups of immigrants. Certain conditions allow for an upward assimilation trajectory where immigrants experience upward socioeconomic mobility and integration into the mainstream society of the host country. But other conditions could result in a less rosy assimilation trajectory where, with each successive generation, immigrants find themselves either stagnating or even experiencing downward class mobility as they are discriminated against or marginalized from mainstream society. I modify and extend this segmented assimilation framework to investigate how the new scientific contexts of return in Asia influence the research productivity and career satisfaction that returning Asian (and non-Asian) scientists experience.

To analyze the postreturn experience, I divide the scientific contexts of return into two components (see also Figure 1.4 in Chapter 1). The first component – which I deal with in this chapter – involves the scientific research system that exists within a country. This system includes the structural conditions, from funding availability to technological infrastructure to administrative support, that allow for research activities to be carried out. The second component – which is dealt with in Chapter 8 – relates to the scientific culture in the research organizations where science is taught and scientific research is conducted.

---

[1] Zhou 1997; Portes and Zhou 1993; Luthra et al. 2018.

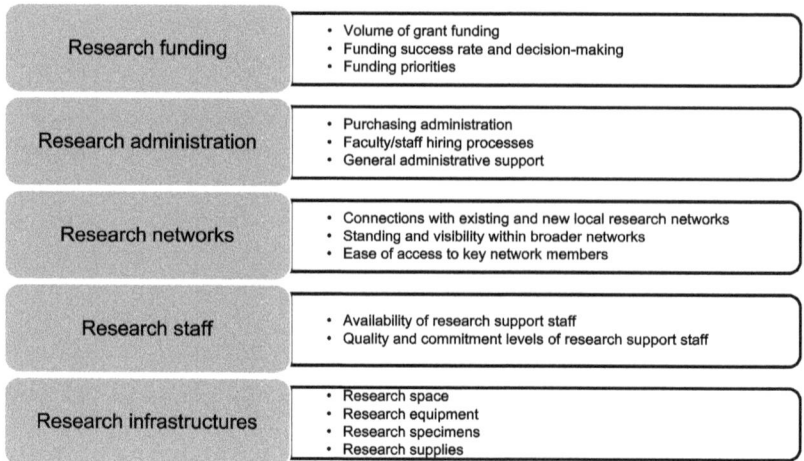

Figure 7.1 Five dimensions of scientific research systems

Together, these two elements determine the range of possibilities for returning scientists who want to conduct high-impact research and nurture new research talent within the country.

To analyze the postreturn experiences of my interviewees, I focus on the five dimensions of Asian scientific research systems that are directly related to scientists' ability to conduct and publish scientific research (see Figure 7.1):

(1) Research funding: the volume of funding available, the rate at which a scientist's grant applications are successful, and the research areas/modalities that are prioritized when allocating funds.
(2) Research administration: the policies and processes surrounding grant applications, fund disbursement and reimbursement, the purchase of research supplies, and also more general administrative functions such as the hiring of personnel and the day-to-day management of the research organization.
(3) Research networks: the degree of connection scientists have, and are able to create, with actual and potential research collaborators and other research peers in their institution, in the same city and country, as well as within the global scientific field, and the ease and frequency of research communications from their location to other network members.

(4) Research staff: the availability of postdoctoral fellows and doctoral, master's and even undergraduate students to work in scientists' labs, and the quality and commitment of these support personnel.
(5) Research infrastructures: the availability and quality of the space, equipment, specimens and supplies available to researchers to conduct their research.

Throughout this chapter, a central question I am interested in answering is: how do the scientific research system characteristics that returnees encounter in their particular Asian country directly or indirectly affect their self-reported research productivity and sense of career fulfilment? I do not make any judgment about which one of these system characteristics is the most important. At the same time, one of my goals is to highlight the system characteristics other than the volume of research funding that are important in influencing scientist productivity. As Suresh, an Indian scientist who left the USA to move to Singapore, shared with me:

> Science is not about money, or it's not *just* about money. Even in India now, when I go, I talk to scientists there and they say, "If I want to buy equipment, if I want to buy consumer goods, I have so much money, I don't know what to do." So people are buying five [units] of a certain kind of microscope for one department when you really need only one, and each microscope costs USD 0.5 million to 1 million. They don't know what to do with the money! Does it mean that their science is as good as the West's? No! It's the people and the institutions [that matter] and that's the same for democracy and the same for the economy. You need the institutions and the systems to make science labs.

The different dimensions of scientific research systems also influence and are influenced by the scientific cultures that develop in a particular research organization. For instance, if a particular research organization has an entrenched and well-known scientific culture of inquiry and egalitarianism, this may encourage more young people to apply for training positions in that organization, increasing the availability of talented and committed research support staff in that setting. However, for the purposes of this chapter, I focus only on the *structures* returnees encountered in the new research organizations they joined in Asia. In the next chapter, I discuss the scientific cultures they also encountered, and how they tried to change some aspects of these cultures. Through these two chapters, I show how change is underway in the contemporary Asian scientific environment. This change is being driven from the top (led by states and the leaders of various research organizations), from the bottom (led by individual researchers focusing on changing the cultures in their labs and classrooms), and also from the outside environment (as Asian societies evolve due to their shifting demographics and growing middle-class populations).

The remainder of this chapter is organized according to each of the five dimensions of scientific research systems. However, it is important to note one key feature of my returnee sample before we begin. While returnees in China, Taiwan and India were primarily native-born scientists, the majority of interviewees in Singapore (80 percent) were not native-born and instead came from other Asian countries (see Table 5.2). I discuss later in this chapter how this foreigner–local spread resulted in somewhat surprising views about the context for scientific research in Singapore, with more negative comments raised about Singapore's research systems than I had expected.

On balance, though, I heard more positive than negative remarks about all four of my Asian country sites. This is partly tied to the fact that returnees were working at elite, well-resourced institutions in their respective countries. Their generally positive comments could also be tied to the fact that, for many returnees, this was their first PI position after completing their postdoctoral training, and so they were buoyed by the excitement of establishing their own lab and research team for the first time. Priya, an Indian scientist who returned to work at one of the top bioscience research institutes in India after a postdoctoral fellowship in Europe, spoke to me about how heady her first year back in India had been. She loved the experience of establishing her own lab and building a team of doctoral students and postdoctoral fellows to drive her research forward:

> I just remember tremendous excitement about coming and setting up my own lab for the first time. It was very hard but it was such an incredible journey to set up a lab and start your own work. You have your own students. It's wonderful to talk to them and engage them. I just enjoyed the whole thing.

Despite these joys, returnees also spoke of encountering institutional roadblocks and unexpected hurdles along the way. Returnees' stories thus balance out the account I gave in Chapter 2 about the ways in which these four Asian countries are investing heavily in bioscience research and revamping their higher education sectors. Returnees' accounts are also able to go deeper than many secondary sources, delving into some of the ongoing challenges each of these countries is experiencing as they seek to fast-track the development of their bioscience sector and raise their profile within the global scientific field. Furthermore, these scientists' stories point to the competition that is heating up between select Asian countries as they seek to differentiate themselves from one another, and recruit from the same pool of returning Asian scientists.

Research Funding

Volume of Funding

Several scientists shared that they had returned to Asia partly because they believed their scientific ambitions were better served in top-flight Asian research universities and institutes, as the funding in these research organizations was now more generous than what was readily available to them in the West. Across the board, all returnees agreed that there was significantly more research funding available in their particular Asian country than what had been available in the past. Scientists who returned to Asia in the 1990s or earlier had lived through these changes and described the difference in the start-up research funding they had received when they returned versus what more recent returnees were being offered. John, who returned to Singapore in the 1990s, recalled how little money he had access to as a junior faculty member when he first came back:

> In those days, when I first joined as a young staff, the resources are scarce and if there's any big money it wouldn't come to me. It goes to someone who is influential, who is a full professor. It won't come to me. So when I put up a research grant proposal, I have to team up with someone senior to help get the money.
>
> Now it is different. Now the young people come in – they are good, good and good! They may get a NRF [National Research Fellowship] and then the NRF will give them USD 3 million to do research. They would have two, three postdocs. Back in those days, we had no postdocs! We had to work [ourselves]. You know, I don't [even] have a lab.

Yogesh, who returned to one of the top universities in India in the early 1990s, also shared that he had received very little by way of start-up research funds, but that the research landscape in India had started improving soon after:

> But then I saw, within a few years, the next batch of [returning] faculty which were recruited after us already got ... start-up funds and it was very well organized. The institute really woke up. They realized that there is something you need to do, something to get the best people in the field. And I think, since then, Indian science has grown exponentially. I think today ... it is a wonderful place to do science. There are so many good institutions that offer good ambience, research ambience, a lot of funding. Government funding has improved significantly. So there are less difficulties in India actually.

Echoing Yogesh's point, more recent returnees in all four case countries noted the increased volume of research funding available in their Asian country. Several interviewees – particularly those who returned to Singapore and China – argued that they were able to access *greater* quantum of funding in Asia as compared to what was available to them in their previous Western country. Deepak, an Indian scientist who trained in the USA before moving to Singapore, noted how getting the volume of funding he needed was easier in Singapore than in the United States, where he felt that funding was drying up:

> Overall, it is working very nicely for me, particularly with the present economic condition. I know that in the US, it is not very easy to get good funding. And particularly the kind of work I do is quite expensive. Because it is a lot of building stuff, and buying big tools, and then large data acquisition, and image analysis. All these. Which is not very easy right now in the US.

Likewise, Rajesh, a senior Indian scientist who had worked in several countries throughout the West and Asia, argued that because of the generous funding available in Singapore, it was the "best place" in the world to conduct research:

> There is no question. Right now, the best place to do science is Singapore, undoubtedly, compared to anywhere in the world. . . . It is the best place to do, the platform to do research. Absolutely no question in my mind.

Funding Rates and Grant Decision-Making

Linked to the greater funding volume, returnees were also generally happy about the grant funding rates in their respective Asian country. They found themselves devoting less time to grant writing because the funding rates were higher than what was prevalent in Western countries. In certain cases, returnees were at national research institutes with dedicated block funding from the central government, which made their research programs easier to support. This was the case with my interviewees at Academia Sinica in Taiwan who acknowledged that the dedicated funding lines they had access to allowed for a more comfortable research existence, compared with researchers at some of Taiwan's universities who had to compete with a much larger pool of applicants for grants. Researchers in Taiwan noted however that funding rates had declined in recent years as more scientists were putting in applications to the National Science Council and the National Health Research Institutes, the two main government funding bodies for biological science research. In Singapore, many returnees were still smarting over the government's reorientation towards more industry-linked and translational research. They also complained that overall funding rates were declining in Singapore as competition heated up between researchers across the country's research universities.

In fact, each Asian country took a slightly different approach to the distribution of central research funds, and each of these systems had its fair share of supporters and detractors. Returnees in Taiwan shared that the communitarian culture of their country encouraged a somewhat even distribution of national research funds so that every researcher received some funds, but no one received a lot. This approach inevitably influenced the size and scale of the research initiatives that individual scientists in Taiwan chose to undertake. Scientists in other countries complained that funding decisions were not made solely on the basis of the scientific merit of their research proposal, but rather were swayed by personal connections (in the case of India and China) or by national priorities (in the case of Singapore).

Several Chinese returnees were concerned that the Chinese research funding system was increasingly weighted towards connections with CCP officials. One returnee in China insisted that the anonymous peer review system widely used in the USA was the gold standard of funding allocation criteria, and that China needed to adopt such a system:

> They[, the Chinese government,] just don't have a system with integrity that can judge grants. You talk to any Chinese scientist, and he would say, "Oh, I don't get money because I don't drink with the members of the funding agency." It's all about who knows who. A system like that can't beat the US. It just cannot.

Liang who had considered returning to either China or Singapore after his postdoctoral training in the USA, decided against working in China because he worried that the only way to win big grants there would be by building strong connections with party officials:

> If you are well-connected, then you are more likely to be successful – even if your proposal is talking about bullshit. I actually visited some of the institutes [before I returned] and immediately, from the very beginning of the conversation, I realized that you are in that problem.
> *[Interviewer:] How did you realize that?*
> Just the kind of conversation that the president of the institute has with me during lunch. He talked to me, "If you can come in [to this institute], I can link you to somebody who is some 'big name' and we can win some grants for you." Something like that. They don't talk about your proposal. They don't know what you're researching about, what you're doing. They just need your title.

Liang worried that if "you're so focused on corruption, then you do not have time to do research." He decided that he preferred the Singapore system, which he deemed to be bureaucratic but not corrupt – an important distinction in his mind – and so accepted an offer from a Singaporean university.

But several other Singapore-based scientists I spoke with were not happy with Singapore's research funding allocation system either. They railed against what they saw as the tyranny of bureaucrats over scientists in funding decisions. Sarkar was one such scientist, who was scathing in his critique of the Singapore funding system:

> So the funding agency wants to ensure a certain outcome, right? But they don't trust anonymous reviewers, who may have their own minds, to ensure the outcome that they want.

So when ... you submit for a big grant, the first-pass screening is done by local bureaucrats, ... based on nonsense grounds I would say, or based on complete incomprehension of the field. Then you get the anonymous reviews. And then it comes back to the bureaucrats for the final decision.

Sarkar was of the opinion that a "rational" system would engage only anonymous scientists as reviewers to score all grant applications and give each a numerical value. In such a system, any grant application that received a cumulative score above a certain threshold would automatically be funded. He could not accept that other criteria, such as the proposal's alignment with a country's national strategic objectives or its commercialization potential, as judged by non-scientists, should influence funding decisions. Sarkar's perspective echoes the idea of a universal science that serves all, rather than the particular needs of any single country. It is also more closely aligned to Mode 1 (basic science) rather than Mode 2 (applied science) thinking, mentioned in Chapter 2. Nonnative Asian scientists who are halfway-returnees may struggle to reconcile their personal research goals with the ambitions of the Asian country they have chosen to work in. But this tension also derives from returning scientists' experience with research funding systems in countries like the United States, where grant monies are dispensed by more independent entities like the National Science Foundation (NSF) and the National Institutes of Health (NIH).

The NSF has clear guidelines for applicants and reviewers about how grant applications should be evaluated. Two criteria – the intellectual merit and the broader impacts of the proposal – are the most important.[2] In assessing the potential impact of a proposed project, reviewers are instructed to consider "the potential to benefit society and contribute to the achievement of specific, desired societal outcomes." Even though the NSF explicitly mentions impacts such as "improved national security [and] increased economic competitiveness of the US" as acceptable,[3] most of the language of the evaluation criteria is framed in universalistic language that belies the fact that

---

[2] "Proposal & Awards Policies & Procedure Guide: Chapter 3 – NSF Proposal Processing and Review," National Science Foundation, NSF 20-1, June 1, 2020, accessed on February 10, 2021, www.nsf.gov/pubs/policydocs/pappg20_1/pappg_3.jsp#IIIA.
[3] "Proposal & Awards Policies & Procedure Guide: Chapter 2 – Proposal Preparation Instructions," National Science Foundation, NSF 20-1, June 1,

Western countries like the USA are also heavily invested in research that will maintain their geopolitical primacy within the global arena.

The NSF process also relies heavily on its pool of over 40,000 external experts in relevant science and engineering fields to evaluate any application on its merits. These reviews are then collated by a NSF program officer who is an expert in the particular scientific field they oversee. The program officer decides on the disbursement of funding monies, with the expectation that they produce a balanced and diverse portfolio of grant awards and limit the amount of political influence in funding decisions.[4]

Other countries will take different approaches in setting up their funding allocation systems, and they could have good reason for including nonexpert civil servants and elected officials in the funding decision. But when scientists move from one research system that uses a particular set of funding rules and norms to another system with a different set of rules, they may bristle at the new (unfamiliar) system, especially if they feel that it is more restrictive. In such situations, it makes sense for countries to do a better job at explaining their funding rationale to newly relocated scientists.

Funding Priorities

Related to the kinds of scientific research that was prioritized in funding decisions, several scientists noted that there was a clear difference between Singapore and the other three Asian countries I studied. They argued that China, India and Taiwan offered more research freedom and flexibility in supporting scientists who pursued basic research. Weihan, who had trained in the USA and had research experience in both Singapore as well as Taiwan, spoke about how differences in funding priorities influenced the ways in which scientists in the two Asian countries were able to pursue particular research agendas:

> There is actually more broad-based research done here [in Taiwan] than there is in Singapore. And I think a big part of it

---

2020, accessed on February 10, 2020, www.nsf.gov/pubs/policydocs/pappg20_1/pappg_2.jsp#IIC2di.

[4] For some wonderful insights into the life of a program officer, read Epstein (2016) and Fisher (2011).

> is because they actually have quite a lot of freedom to do whatever they want. When the funding agency gives money, they don't really require that you have to produce anything applied or translational as a big part of it. So there is a lot more basic science that can be done here [in Taiwan] and so more scientists here are able to explore different avenues. And that's, in a way, where creativity comes in, you know? You're not trying to produce something; just going along with what [you] see and ... from there, suddenly, you discover something.

Weihan's belief that Taiwan offered relatively greater support for basic research is likely linked to the fact that he worked at Academia Sinica. Taiwan's national academy continues to exert significant clout in the Taiwanese government's long-term scientific planning process and has separate funding lines, giving its researchers more research independence. But Weihan was also speaking to how basic research funding in Taiwan was not as tightly tied to the need to publish a fixed number of papers or produce other tangible results by the end of the project.

Returnees in the two larger Asian countries in my sample – China and India – also found their countries to be more broad-based in terms of their relative funding priorities between applied and basic research. Yuewang, who had worked in both China and Singapore, noted that Chinese universities' fundamental priority was to increase their research standing by having more publications in top journals and more citation counts. This ironically resulted in a certain degree of openness about research topics in China:

> I think China is actually doing a better job than Singapore. Because Singapore is so conscious all the time about the economic drive behind science. In China, as long as you get good publications in terms of good journals, people will not be so bothered by saying you are doing ecology on this very arcane species. People will respect you regardless of what you are working on. But [in Singapore,] it's too practical driven. Science in China is a bit more liberal.

Scientists whose research does not have obvious translational applications that the state would deem relevant, or if they are not interested in developing their research in this direction, may therefore struggle in Singapore. For instance, interviewees told me that, in the 2000s, Singapore began moving away from funding basic research in plant biology, given the country's lack of a large agricultural sector.[5] Asian scientists who work in this area may feel disinclined to return, or (if they are already in Singapore) may choose to leave the country because they do not see their research specialization being valued by the state. I interviewed several Singapore-based researchers in plant biology who were seeking jobs elsewhere because they did not see a viable future for their research interests in Singapore.

There is an open question as to the long-term impact of this fixation on commercially oriented research results. Returnees repeatedly emphasized the need for governments to accept a longer time horizon when it came to research investments, but not all countries have the appetite for patience in this regard. Wai Ming complained that Singapore's government worked on a very short time horizon with which to produce commercially viable research results, thus preventing the decades-long process of building deep expertise in a subject area:

> So the government today thinks that "water is important" and so they pump money into water [research]. Then they say, "I don't want water [research]." Next day, they say, "Energy is more important, let's do clean energy [research]." Then they pump in our money into clean energy research. Let's have the former water expert – you become an energy expert. But you know, a rolling stone cannot gather moss.

Returned scientists talked about how these differing funding priorities influenced the competition between their university/institute and those in other countries when it came to recruiting Asian researchers who were interested in returning to Asia. Often, the bargaining chip in their recruitment strategy was the volume of research funding available to new hires, but a country's funding priorities also made an important difference. Jun Wei, a senior scientist in Singapore, shared how Singapore's smallness and reputation for prioritizing translational research sometimes worked against them during their recruitment efforts:

> Now, we have to compete with China. And it's not really pleasant, the competition. [*Jun Wei laughs*] Because, for the last two

---

[5] This shift may be reversing however as Singapore subsequently set itself the goal of producing 30 percent of its food needs by 2030 ("Food Farming," Singapore Food Agency, updated February 17, 2021, accessed on February 27, 2021, www.sfa.gov.sg/food-farming).

times, we actually failed to recruit. ... I think, for Singapore, there are a few things people will view: it's a small community. There is nothing to do about that, right? And we have a little bit of negative image out there at the moment, because there is such a strong push of industrial alignment strategies. ... For many young people, they feel like "Oh, [researchers in Singapore] are limited, structured, organized." And that's not really the most favorite environment for research. Because they want to have total freedom, high level of flexibility, with stable funding. That's what they are looking for.

Research Administration

A separate but equally important domain of scientific research systems relates to the day-to-day management of and bureaucracy surrounding grant applications, grant management, the ordering of research supplies, staff hiring procedures and more general administrative functions, both within returnees' research organizations and at the national level. Across the board, returnees spoke of encountering more administrative hurdles and bureaucratic inefficiencies in their new Asian places of work compared with their workplaces in the West (and particularly the USA). All returnees agreed that this led to regrettable delays in their research productivity and to feelings of frustration.

I argue that these (presumed) system flaws stemmed from differing priorities for the administrative work processes that support research endeavors, with the efficiency (and speed) of these processes losing out against the need for accountability and financial prudence. In recent decades, there has been significant scholarship on the increasing rationalization of (primarily Western) universities' administration and governance processes.[6] Critics label this trend the "corporatization" of higher education and complain about the detrimental effects that these university reform processes have had on internal governance structures, budgets and organizational culture, even as these reforms were linked to calls for increased accountability to the wider public of students and also taxpayers. Christensen notes that these reform processes have been framed as efforts to increase efficiency and accountability, but that "it has never been quite clear what is really meant by

[6] Ramirez 2010; Ferlie et al. 2008; Amaral 2008.

efficiency in an institution like a university, for example, with respect to research activities" (2011:508).

Purchasing Administration

Given this backdrop, it was fascinating listening to my interviewees speak about how they found university administrative systems in their Western universities to be *more* efficient than the ones in their new Asian places of work. An important rider to note here is that all the institutions in Asia where interviewees trained, and then subsequently returned to, were public. This was the case even in countries (like Singapore) where the national universities have been made more autonomous in their internal decision-making.[7] In contrast, the majority of my interviewees had completed their postdoctoral training in top private universities in the USA. Perhaps because Asian institutions saw themselves as dispersing and managing public funds to support scientific research for the national good, they were motivated by different goals and performance indicators compared to the administrators at elite, well-endowed private institutions in the United States, for instance, who were less concerned about accounting for every dollar spent. Most Asian universities emphasized accountability in order to prevent corruption, embezzlement or improper use of department funds, and this led to a preponderance of forms, multiple approval levels and other bureaucratic interventions. In contrast, American research universities appeared to be more focused on efficiency and speed. European universities were situated somewhere in the middle of this spectrum.

Rajesh tried to explain to me, in a humorous way, the differing worldviews of scientists versus the finance staff at his Singapore research university:

> I think one of the things that terrifies a biological scientist is the red tape. I think part of that comes from [the fact that] scientists are probably by far the laziest people in terms of doing paperwork, right? [So] I am a bit sympathetic towards somebody who says, "You have to fill these three forms to get this chemical." Because accountants couldn't care less. They don't want to punish us. It is just that when the auditor general sends

---

[7] Ng and Tan 2010; Lee and Gopinathan 2008.

somebody to say, "Well, they spent $5 more here," which when everybody does it over five years it's a $5 million loss to the government. And then, the public is going to lynch them and say, "Why did you waste money?" So to make that not happen, they need these three forms to be filled. And if I look at this [university's] procurement system, [it is clear that] nobody trusts nobody.

Between efficiency and accountability, interviewees were overwhelmingly in favor of administrative processes that prioritized the former. Yong Kai, a Taiwanese microbiologist, described his surprise at how quickly his American doctoral department handled purchase orders, equipment requisition requests and other administrative support processes:

I was really impressed about the efficiency [in the USA]. In Taiwan, and also in Asian countries, ... most of the countries have a lot of bureaucracy. If you want, for example in Taiwan, on this campus, if you want to purchase an instrument, if it is very expensive, you have to go through a lot of procedures. Because they want to try to prevent you from corruption. So you need to go through a lot of steps and a lot of questions. But in the US, it's very straightforward. Probably because it was a private institution [that I went to in the USA], so it also avoids the government bureaucracy. But I was really amazed by the efficiency. Doesn't matter how much it takes, they just want to know if you get a good result. I was so impressed.

As can be seen from his quote above, Yong Kai attributed the efficiency of his American doctoral institution to the fact that it was a private university, while explaining the bureaucratic checks in Asian institutions as their attempt to squash corruption. But other scientists (particularly from India) argued that the many layers of approval required for any purchase only enabled *more* corruption. Indian scientists complained about the additional budgets they needed to set aside when purchasing foreign equipment or supplies because of the onerous tariff structure imposed by Indian customs agencies. Many of them assumed that some of these extra fees were being pocketed by corrupt officials comfortably ensconced within India's complicated import approval structure. In other countries, scientists complained that the only way to circumvent a system of multiple approvals and long wait times was to leverage your rank. Liu, from China, lamented that the bureaucratic apparatus in China moved faster for senior scientists or those who were politically well-connected: "Unfortunately, you need to demonstrate you're powerful before you receive anything."

Scientists were quick to highlight the negative impact of such bureaucratic delays in purchasing research supplies, arguing that it prevented them from pursuing spontaneous research ideas. In this regard, India was the worst culprit by far among my four Asian case studies. All of the returned Indian scientists I interviewed lamented the inefficient and, at times dysfunctional, administrative processes in their workplaces in India that lowered their research productivity. Dodi, who returned to India after completing his postdoctoral training in the USA, complained about how much longer it now took him to receive research supplies, as compared to when he had worked in the USA:

> This was not the case when I was at MIT. If there is some chemical I require, before going out [at the end of the day], we'll just send that order. Next day morning, before I come to the lab, it is delivered. ... In the Indian system, if you want to work, ... you have to plan "a little" in advance. You can't do last-minute things.

Likewise, Manas remarked that when the research support environment was efficient and smooth-running, it made conducting research "stress free." But Manas thought the Indian research environment was far from reaching this state of efficiency:

> [In the USA,] you get a lot of time for your work, other than for running around. In a day, if you have eight hours, probably for eight hours you can think about your work. But in India, if you have eight hours, [only] two hours you are actually working. [For] six hours, you have a lot of other issues going on!

Manas found it embarrassing that even though India was renowned as a source of skilled software engineers for the global economy, its scientific institutions had not automated, streamlined or moved more of their research support processes online. He argued that it was demotivating for researchers in India to have to spend so much time on administrative tasks, leading to them lowering their research ambitions so as to avoid the unnecessary stress or deciding to leave.

While not as dire as the situation in India, even an advanced economy like Singapore was not immune to such criticism. A Singaporean scientist, Sandra, shared "how hard it is for science to advance when you're so sticky about rules." Another returnee in Singapore, Jonas, complained how there were many more checks, and therefore also more delays, when ordering research supplies in Singapore, compared to the process he encountered during his US postdoctoral fellowship:

> In my old lab [in the USA], I order something, I got it the next day, no questions asked. [*Jonas laughs*] No one will ever ask you any questions. Or ... you just tell that person, "Hey, I want to order this." They'll say, "Okay. You give me a quote, I will order for you."
>
> Here [in Singapore] you have to go through multiple layers. You go to the finance [department] and then finance has to generate a PO [purchase order] for you. And then it takes a long time to get anything.

Jonas shared that his colleagues who had been in Singapore longer than him had advised him to "just learn to be very patient." Likewise, Manas pegged Singapore as being somewhere between India and the USA in terms of the efficiency of its systems, but he attributed Singapore's slow processing times to both its layered bureaucracy but also the smallness of the country, which meant that almost all research supplies had to be imported.

Another Singapore-based Indian scientist, Deepak, said he felt hemmed in by his Singapore workplace's purchasing and tender rules, which prioritized lowering costs over other principles such as product quality:

> Suppose I want to buy a laser. So many Chinese companies will give you a quotation which is very cheap. But I know this [product of theirs] won't work, because it won't be stable after one month. So there's two sides of requirements for lasers. One is precision balance and one is the cost. For precision balance, I have to go for some good companies, because that is the key thing. If the laser is not stable, then with time, it is of no use. Then the whole money is a waste.

Deepak complained that the purchasing rules in his research institution in Singapore were rigidly geared towards reducing costs, requiring him to spend a lot of time justifying why he wanted to purchase a needed piece of equipment from a vendor that did not submit the cheapest bid. Scientists like Deepak were not arguing for the removal of all oversight on their spending or research. Rather, they complained that the administrative and research sides of their Asian places of work were often at cross-purposes, thereby slowing down their research progress.

Hiring and Other Administrative Functions

Scientists in India also complained about the long delays in faculty hiring – a problem that seems largely limited to India. Speaking from firsthand experience, Karthik (introduced at the start of this chapter) recalled waiting one-and-a-half years from the submission of his job application before receiving a formal job offer from one of the top bioscience research institutes in the country. Another institute in India took a year to make him an offer. Karthik shared, "I had forgotten how slowly things move in India. ... After one year, they kept saying, 'Yes, yes, you have the job, don't worry about it. But you know you can never trust this until you have it in writing and so that took six more months." Luckily, Karthik's postdoctoral supervisor in the USA kept him on for an extra year while Karthik waited to hear back from the Indian organizations he had applied to. While Karthik was committed to returning, such administrative nuisances can deter other overseas-based Indian scientists from returning to India.

Research organizations in Singapore took a very different approach. Especially during the early days of the country's big push into bioscience research, when it was setting up its Biopolis campus, the directors of its life science-related research institutes were on an aggressive recruitment drive. A Chinese scientist, who had been recruited to join one of the research institutes at Biopolis in the early 2000s after completing his postdoctoral training in the USA, told me the pitch he had been given when he met with the new director of the institute:

> He basically told me what the institution is going to look like, what kind of funding we are going to have. And there is a humongous amount of money and the guarantee that the government will give five years of funding money. Well, we could do a lot of things there with this money. That is really

the main factor for me, if you like. ... So I said, let's go to Singapore! So that's the story how I ended up here.

This scientist went on to talk about his former director's recruitment strategy, which he felt had helped their institute become very successful:

> So he basically recruited mainly the small fishes like us – a lot of them. That's his philosophy. And I think that turned out pretty good. Because [our research institute] is now considered one of the most successful institutes [in Singapore] in terms of productivity, publications and all this. Because we are all young and we are all coming in hungry. Hungry for success and hungry for whatever. And you give us money – and we never have so much money before – so we run [with it]! I guess through the process, we all grew up and are mature now. And many of us have become famous, not only in Singapore but internationally, as the leaders. But, during the process, [our institute] also grows. So I think that philosophy has worked out very well ... and that's why we still continue that philosophy. We still continue investing in the young scientists and at this moment we still recruit quite a few people, very young people. And I think that's the driver, the engine, the real big engine.

Returnees in China also emphasized how much of a push the Chinese government is making to woo talented Chinese scientists back to China:

> But yeah, so China has been quite aggressive in terms of recruiting promising candidates back. And every year there is at least two rounds of big recruitment from the States. This is the [central] government and then, every university, they have their own campaigns and the Chinese Academy [CAS] also.

Meanwhile, research organizations in Taiwan have realized the need for a more streamlined and open decision-making process for recruiting talent. Akemi, a Taiwanese molecular biologist who had returned to start work in a new research institute in Taiwan in the early 2000s, spoke of how hiring priorities and decisions were now made in her institute with everyone being given an equal say. Akemi saw this approach as the adoption of Western (and specifically American) governance ideals of transparency and efficiency:

> So our institute, from the very beginning, is the American style and now, more and more institutes [in Taiwan] are like us. I think it's changing for the better. For example, we have faculty meetings. The committee will decide whether this year, in the [faculty hiring] advertisement, we want a special kind of person – a special kind of background – or we just want to recruit the best people in molecular biology. And then we interview them and then everybody votes, and then when we finally decide to recruit them, we will give them a good start-up [fund], a good lab space and we won't let them teach or do any other administrative work for the first few years. So it's a good research environment. I wouldn't say that for every institute in Taiwan but I think they are changing toward that. I think, finally, everybody realized that that's the way to compete. I mean everybody wants to become a better institute.

As Akemi noted, Asian research organizations that wanted to be competitive in hiring the best local and overseas faculty needed to work not only on front-end issues like funding but also on backend administrative and governance processes.

From my interviews, it was clear that Indian research organizations had the most room to improve when it came to their general administrative processes. Perhaps the worst story I was told about the dysfunction within the administrative support systems of research organizations in India came from Manu, a scientist who had returned to India but then became so frustrated after a few years that he started looking for jobs back in the USA. He gave me the example of how his Indian institute's research director had to personally manage the budget and operations of the institute's staff canteen:

> The [administrative] support infrastructure is absent. The director of this institution deals with things like canteen food cost and events. The head of an institution should not have to do this! The reason they did what they did is because they know very well that the infrastructure is absent. And if they don't do it, things will go downhill very quickly. So they're doing it out of desperation. Otherwise, things wouldn't work. But that hurts us in terms of productivity.

Manu was adamant that it was this kind of wasted time at all levels of the national scientific establishment that continues to slow down Indian research.

Research Networks

Another area where significant heterogeneity existed across the four Asian case countries regarded the ease with which returnees in each country were able to connect with a network of scientists working on similar topics and using similar methodologies. This network might operate within their new research organization or in the same city/region within their particular Asian country, but the degree of connection to potential collaborators and role models in the West was also important.

Returnees gave mixed responses to my questions about their new networks in Asia. On the one hand, many were excited that increasing return numbers meant that the local scientific community in their subfield was approaching a critical mass. They talked about their own role in building *new* Asia-based networks and finding great fulfilment in playing a central role in these new research communities. On the other hand, they also acknowledged that the core community in their respective subfield was still centered in the West, resulting in them feeling somewhat excluded from those networks, and steadily losing status if they did not actively work to stay visible in those circles. Depending upon how globally ambitious an interviewee was, this reduced centrality within the international scientific network could be more or less painful.

Building Networks

Of all the returnees I interviewed, the ecologists I spoke to in Bangalore (known officially as Bengaluru) were the most excited about the growing research community they were now a part of in India. One such ecologist Ajit spoke about the recent emergence of what he called an "ecosystem" of like-minded scholars spread across various institutes in Bangalore and how energizing it was to be part of this group. While Ajit had always wanted to return to India despite the known research hurdles, he felt even more convinced of the rightness of his decision because of how enmeshed he now was within a lively community of young researchers in Bangalore:

> We certainly love Bangalore. We don't want to move from Bangalore, because Bangalore is actually the place to be for ecology [in India] because of the concentration of ecologists there. ... It's got several research institutes. It has both

government and private. ... What's happened is, in the last ten years, a lot of Indian researchers like myself and a whole bunch of ... young guys who had gone to the US and elsewhere have gotten their PhDs at some of the top labs or universities in the world and have come back with that cutting-edge expertise. ... It has taken a long time to get to this stage, but now we have a critical mass. So now, we are training our own PhD students in the same rigorous manner. And our PhD students are doing really well. They're publishing really well, because if you look at the trajectory of publications by Indian ecologists, for years and years and years, you rarely find papers in top international journals. But if you look in the last five, ten years, it's just exploding. ... Finally, it means we are coming of age in terms of contributing through [our local] ecological knowledge.

Bangalore is an increasingly well-known and trusted location within the overseas research community as a center for high-quality biological and ecological research.[8] Returnees shared that international scholars were regularly coming through the city to give talks or attend seminars, making international collaborations easier.[9] Ajit was of the opinion that, despite being back in India, he was "very much linked in with the international conservation network."

Being located in what could be considered an "Asian node" within the global research network was viewed as a key strategy to maintaining ties with the international scholarly community. As I mentioned in Chapter 5, the return decision often hinged on which city or research institution, in a particular country, Asian scientists were able to return to. This question was not so much of a concern in Singapore given its small size and the high standing of its research universities. But for scientists considering a return to countries like China and India, it made a difference if they were going to relocate to a city like Beijing or Bangalore, or to more remote parts of their respective country. Not only were the research resources better in these more prominent cities, but institutions in these locations were more visible to researchers outside the country. Returnees in Taipei, Singapore and Beijing all spoke of inviting international researchers to their institutions as there was more funding being made available to them to organize

---

[8] Department of Biotechnology 2012; Van Noorden 2015.
[9] When I visited different life science research centers in Bangalore, I also noticed posters for various upcoming talks by visiting foreign faculty.

international conferences and workshops, and to fly in overseas participants for local events. Being located in an Asian city that foreign researchers were curious about visiting as tourists also helped in this regard. Su-Wei from Taiwan noted that, rather than trying to attend as many international conferences in the West as possible, she organized international conferences in her own institution. Su-Wei shared that the director of her research institute was always supportive of her efforts in this regard and that inviting foreign scholars to visit Taiwan for a week or so was relatively easy:

> If I want to organize a symposium related to proteins, it's okay. I just talk to my director. I then go find another PI in another institute in a related field, and each one finds money in their institute to sponsor one speaker. So then, we could organize a symposium with about half foreign speakers and half local speakers. ... This, in fact, is better than me going out to an international symposium where I don't know anyone.

Returnees were also instrumental in creating transnational connections between their new Asian place of work and the research organization in the West where they previously trained. They became a conduit for students and trainees from their Asian institution to find short- or long-term placements at the Western organization. This helped them maintain their international research connections, but it also funneled their Asian students into Western institutions more easily. Feng Cheng, a Taiwanese scientist who had trained in Canada and then returned to teach in Taiwan, established a cooperation agreement with his doctoral university in Canada to allow for a graduate student exchange program in the summers:

> I think before I went to that program only two Taiwanese alumni from my former department went to [the Canadian university]. But after I went to that program, I introduced them to [my Taiwanese department]. ... So after I came back as an assistant professor, I started to have a kind of cooperation agreement with [the Canadian university]. So now we specified the agreement, so we're going to have a student exchange program, a summer program or a graduate student exchange program. So after I came back, I just promote that.

As with earlier generations of Western-trained Asian scientists, returnees continue to support the brain circulations of their students. But the nature of these circulations are looking increasingly different, becoming more lateral and bidirectional.

Absence from Networks

Even as they worked hard to build these transnational networks for themselves and their students, most returnees agreed that being based in Asia was still a handicap. This was in large part because it was harder for them to attend as many of the same conferences and workshops as Western-based scientists. These were the venues where researchers networked with other scholars, learned about each other's work-in-progress and established new connections and potential research collaborations. Sneha, an Indian neuroscientist, was adamant that being outside her network's core (which she saw as situated in the USA) was detrimental to her research:

> It's lonely and you don't have a critical mass in your area of science yet. It's growing [in India], but it's still nowhere near the critical mass in the US. So you have to travel a lot so that you don't end up disconnected from the rest of the world in terms of science. ... There is no question there is a disadvantage. Absolutely. It's like being asked to run a marathon but you have a little ball around your ankle.

While Sneha acknowledged that the situation in India for neuroscience research had improved in recent years, she still felt frustrated that returning to India meant she was no longer in her network's core. Sneha also noted that increasing her international travel could not fully compensate for her semiperipheral network position. Unlike the Indian ecologists I interviewed, neuroscience in India did not yet have significant representation on the world stage.

Akemi, a senior scientist who had returned to Taiwan as an assistant professor and then risen through the ranks, also talked about the ongoing disadvantages of being based in Taiwan versus the advantages of being based in the West, and especially in the USA:

> They can get together for small conferences but we can't. And they get to review each other's grants and they get to talk to each other more. ... I know people in my own very specialized field but, outside of my field, I don't know that many people. But they

do because they run into people in grant review panels, in meetings. I think that's the drawback, yeah. ... In the United States, they are simply bigger and have more people.

Akemi's explanation raises an important point about the role that grant funding institutions like the NSF play in supporting national-level research networks by calling all the grant awardees within a certain program for annual meetings. The same argument was raised by Murali, a US-trained scientist who returned to India and then eventually decided to leave after three years and return to the USA. Murali noted that, while in India, he had needed to undertake a "ridiculous" amount of international travel to stay connected to his overseas networks. But all this time spent travelling meant he was not in his lab conducting research and also that he was away from his family. Murali also spoke about the role played by grant-funding bodies in fostering regional networks in Europe and the USA:

> Conferences are a very small part of the peer group in the US or in Europe. ... For example, if you are in Europe, then you submit for a European grant. You apply and if you succeed in that grant, you then become a reviewer for other grants and, in review panels, you meet your peers. [In these panels,] you are not always going to just talk about that specific review process, you also get to be sharing about your own science. The same happens to a much greater degree in the US, with NSF and NIH.

Gender played an important role in the development and maintenance of networks as well, compounding the existing peripherality experienced by women scientists based in Asia (as I discussed in the previous chapter). Su-Wei, in Taiwan, said she only travelled internationally for conferences once a year. These trips were usually to the United States, but also sometimes to the UK. She explained that part of the reason why she did not travel as much was because her husband did not like her traveling on her own and because it was too expensive to bring her whole family with her for such a long visit:

> Once a year is enough, because my expense is covered by my institute but for my big family including my mother-in-law, maybe I need to spend NTD 0.4 million. That's big money. So okay, once per year is enough.

In this manner, women scientists I interviewed in Asia – particularly women who were mothers – suffered the most in terms of their access to international research networks.

Losing Status

Separate from network *access* issues, returnees also argued that there were structural biases built into the global scientific field that privileged research produced by Western research organizations, over work from other parts of the world. For scientists who had flourishing careers in the West before their return to Asia, they attributed the drop in their article acceptance rates after their return to Asia to the lower status of Asian institutions and, by extension, Asia-based scientists, in the global scientific field. Annya, a biochemist in Taiwan, shared how demoralizing she found this, because it signaled to her that her research was not being judged on its merits but rather on the basis of how well-known she was and where she was based:

> Sometimes, the connection is very important. ... In real science, it shouldn't be important. The most important thing should be what you have contributed, and what you have discovered. But often that same kind of concept, the same package, the same story [from you], cannot be published in a high-end journal. But you will read from the articles [that were published] in the high-end journal, [those authors] also do this, and we can also do that, but we never get in that [journal]. Because all the papers are going through a review system, and it depends on who reviews your papers. So say, if I noticed this lab [submitted a paper] and the author is my friend, I wouldn't give very harsh comments. I will try to help and I will say, okay, you can revise this and that, and then they get it [accepted]. This is really true. Because previously our former vice president, he was a professor in [a public university in the USA] and he published lots of *Nature* papers. But he has no idea why, when he moved back to Taiwan, he suddenly cannot publish any more articles in those journals. So I think any system run by humans will have this kind of problem.

Likewise, Yogesh from India surmised that, after moving to Asia, his research was not as "noticed" as it would have been if he had stayed in the USA and worked on the same research projects. He specifically mentioned that "it is harder to publish from India in the top-level journals." Meanwhile interviewees in China complained against what they perceived as the unfair reputation that Asian scholars from Asian institutions produced shoddy or questionable research (highlighted in Chapter 2 as well). Li Qiang felt that studies conducted by China-based scientists were unjustly given a bad reputation around questions of their reproducibility:

> If anything, 90 percent of *all* research papers cannot be replicated, not only [those from] China. Compared to other researchers, I feel Chinese people are actually the careful persons in the research area. I think it's a little bit of bias actually.

Returned scientists were thus very aware of the social embeddedness of scientific research and how they had to work harder, and conduct more confirmation tests, in order to remove any doubt in the minds of reviewers at the top journals in their subfield about the validity of their results from Asia.

Research Staff

Another key dimension of a functioning research system is the availability of good quality postdoctoral fellows, doctoral and master's level students and even undergraduate students, who can serve as research hands in a scientist's lab. In all four Asian countries, returnees who worked at research organizations that had in-house doctoral programs, appreciated being automatically assigned one to three graduate students to work in their labs without having to pay for them out of their start-up funds. Li Jie, a Chinese scientist who moved to Singapore, shared that, in addition to the plentiful research funds available to him at his university, "if you come here, the graduate students are free. At the beginning [of our contracts], we were guaranteed two or three graduates for free. Later, it depends, of course, if you have grants. You can use your grant to support them. But if you don't have grants, sometimes you [still] get students for free." In India, meanwhile, scientists at top universities and research institutes with degree-granting authority remarked that they had access to an "unlimited" number of graduate students.

However, returnees – especially those in research institutes that did not have access to doctoral students – uniformly complained about the shortage of *postdoctoral* trainees who would have been even more useful to them. Given that postdoctoral trainees had successfully completed their doctoral training, these individuals tend to be able to conduct more independent research, require less mentoring, and can even take on some of the supervisory work in the lab. But because of the ongoing appeal of the West – and particularly the USA – as the destination of choice for postdoctoral training, returning Asian scientists found themselves hampered by the dearth of top-notch postdoctoral trainees. Rajesh, the Indian scientist now working in Singapore, argued that this manpower issue was the only significant hurdle holding back the country:

> The problem we face is recruiting the peons of science, the ones that actually do science at the bench. The charm is still the US, even though there are so many negatives about it. So the best postdocs from Europe, from India, from Asian countries, from Singapore itself, they want to go to the US. Even though in some respects, we are doing much better science than what is happening in our counterparts, they still want to go [there]. So we lose a lot of talent. We don't attract the best talent.

Scientists I interviewed in Asia were aware that, while returning to Asia had made sense for their own careers, it was harder to convince their top doctoral students to stay in Asia and work with them, rather than pursue postdoctoral training in the West. As another Singapore-based scientist put it: "The top students from here don't want to be here [in Singapore]; they want to go places," demonstrating that Singapore was not yet viewed as a global hub for research *training* excellence that would draw the best postdoctoral applicants from around the world. In other cases, interviewees in Asia highlighted push factors in their home country that further dampened the quality of postdoctoral applicants they received. Despite growing complaints about postdoctoral training systems in Western countries,[10] interviewees believed that the problems with Asian postdoctoral training systems were far worse. (I have already discussed some of these problems in Chapter 3.) One issue raised repeatedly was the hierarchical nature of postdoctoral systems in Asia as compared to those in the West. Meanwhile in the case of

---

[10] NRC 2009, 2000; Mitchell et al. 2013; Nuffield Council on Bioethics 2014; Alexander von Humboldt Foundation 2013; Dillon 2003; McDowell et al. 2015.

India, there was a general consensus among returnees that the Indian postdoctoral training system had room to improve on several fronts, from salary and benefits to prestige.

Asian countries were actively establishing more postdoctoral training programs to encourage Asian PhD holders to pursue their postdoctoral training in Asia itself, but given the scientific cultural capital that could be acquired from spending time in the West, and the reward systems in Asian countries that privileged those with Western postdoctoral training and connections, it was still difficult to entice top Asian students to choose to stay in Asia. However, intra-Asian mobility for postdoctoral training was on the rise. Wan Lei, a Chinese scientist who chose to return to work in Singapore, shared that Indian and Chinese doctoral students were increasingly considering a postdoctoral stint in Singapore as a viable alternative to the USA. The issue in Singapore was not so much the availability of postdoctoral applicants but rather the quality of these trainees. According to Wan Lei, Singapore was still seen as a Plan B by the very best Asian science students:

> Because compared with the US, it's more difficult to have good students [here in Singapore]. Especially from other countries, such as from India or China. I think that, for example, for Chinese students, some of them choose Singapore because the requirement for English is a little bit lower here. But for Indians, you know, usually Indians have better English [language] levels, so for any good student from India, maybe they just choose the United States, right? ... Because I tried to recruit some students from both India and China, and I found that problem. Because usually, they just want you to wait. They want first to check the situation from the US. If they couldn't [get a position there], only then maybe they will accept the offer from you.

Similar observations were made by returnees in other countries who noted that, even when they could recruit postdoctoral trainees, the best applicants still looked to Western countries, and particularly the USA, for fellowship positions. Vanya, in India, lamented about how she invested so much of her time training her strongest students during their doctoral studies, only to see them leave after they completed their PhD:

Now, we train them. They do their PhD [in India] and then they go for their postdocs abroad. So we feel the pinch because we are not able to see the advantage the US has when they can attract postdocs from all over the world. The best postdocs from all over the world want to go to those labs. We are not yet able to attract the best postdocs from all over the world to come to India. That hasn't even begun. I mean, it's far from it. And that will take a period of time. ... And so, that's just the realization that you have, that we are not on par at all.

The only exception to this rule were the Singaporean doctoral scholarship holders who were required to return to Singapore, after completing their doctoral training, to serve out a postdoctoral training period in a research institute in Singapore. Interviewees in Singapore raved about the quality of these trainees, but they also noted these scholars' excellence made scientists reluctant to hire other postdoctoral trainees who were often not of the same calibre. One scientist in Singapore highlighted how this divergence in quality led him to hire fewer non-Singaporean postdoctoral scholars, thus reducing the diversity in his lab:

> We rely on the A*STAR scholars a lot. But we are losing something that's very important, which is diversity. The institute looks much less diverse than before. It can be good and it can be bad. I will say the foreign postdocs are coming less and less. Because the A*STAR scholars are coming, and you know, they are actually very good. They make the standard so high. If I had three scholars in my group, when I have [a non-scholar] postdoc come in to apply, I look at them, and well no.
>
> *[Interviewer:] So they are crowding out the others?*
>
> Because they have set the standards so high. ... Before it was okay. I will go and take whomever comes here, right? Now we say, "Well, why should I take this guy? Look at what we've got so far in the lab." And so, our diversity is definitely going down now. ... For research, that is not the best environment. You would like to encourage diversity at least for the research environment and that's very important.

Thus, even when scientists in Singapore were able to recruit top-notch returning Singaporean scholars as postdoctoral trainees, there were still issues they faced when it came to building a diverse research team. I will talk more in Chapter 8 about how returning scientists were trying to increase the diversity of their labs, and also about the challenges with managing diversity in Asia.

Research Infrastructures

The last dimension of scientific research systems that can influence scientists' productivity are the technological and other research-related infrastructures available to them in their place of work. These infrastructures include laboratory space, equipment, specimens and supplies. This was another area where there was discrepancy across my four Asian countries, with India once again bringing up the rear.

Manojit, who first went back to India after completing his PhD in the USA, but left after a couple of years and moved to Singapore, spoke about the poor state of the technological infrastructure and research facilities in India as part of the reason why he decided to leave:

> For my work, I have to depend on many people. I work on small animal imaging, so I have to depend on an animal facility. I work with lasers and all other stuff, which needs at least a clean environment. Sometimes I need to work with clinical people. So I am always dependent on others. For me to survive or do well in that system [in India] is very challenging. And so, I thought, it's not the best idea for an experimentalist to be in India.

Manojit insisted that India's technological infrastructure was nowhere near to where it needed to be to truly compete with countries in the West. For theoreticians who needed only graduate students and a "pen and paper" to do their work, Manojit thought India was a great place to be a researcher. But he was convinced that, for experimentalists, it required too much effort to be based in India.

In contrast, returnees in Taiwan boasted about the state of the technology that was now available to them in the country. Runchen noted that, in the past, certain individual scientists at her institution would hoard their equipment such that the only way to access this equipment for one's experiments was to collaborate with these scientists, or purchase one's own equipment, leading to high levels of redundancy. But her current institute had decided to establish core facilities that everyone in the institute and also elsewhere in Taiwan could access:

> We have six mass spectrometry machines. The best ones! And the price that we charge is much lower and that's why professors come here. They are shocked about the price we charge. ... [And] we have lots of people with expertise in that equipment. They will help teach you how to handle your sample, how to do the experiment. They could do it for you or you can have your student go there and they teach them how to do that. They also offer a lot of courses.

According to Runchen, the mark of a good research environment was the availability of this type of centralized core facilities that could be enjoyed by multiple faculty.

Likewise, Singapore has invested heavily in building up its core research facilities, though it is somewhat hampered by its small size and so has to be judicious in deciding the particular facilities to construct and the equipment to purchase that would have a high enough utilization rate so as to justify the expense. Haolie, a senior scientist I interviewed in Singapore, explained to me how the country had invested in a synchrotron, which is an essential but also very expensive research facility that uses high-energy radiation to look inside living organisms.[11] While it was a big step forward for the life science research community in Singapore when the Singapore Synchrotron Light Source was established in 1999,[12] the scientist noted that Singapore had invested in a relatively small synchrotron compared to the Taiwan Photon Source which is one of the brightest synchronized light sources in the world, set up as part of the National Synchrotron Radiation Research Center in the NHRI in Hsinchu.[13]

An important point raised by Hongyu, who had trained in both the UK and the USA before returning to China, was how such cross-national variations in research infrastructures existed even within the West. He argued that US research organizations offered a better set of technological resources to researchers compared to those in the UK. He had first observed this during his training years, but felt that this was still the case. Speaking about the UK, he noted how, as a doctoral

---

[11] Synchrotrons are also used to see inside advanced nano-engineering equipment.
[12] "Singapore Synchrotron Light Source," National University of Singapore, last modified July 16, 2010, accessed on February 27, 2021, http://ssls.nus.edu.sg/about_us/aboutus.html.
[13] "About NSRRC," National Synchrotron Radiation Research Center, accessed on February 27, 2021, www.nsrrc.org.tw/english/about_1.aspx.

student at one of the top universities in the UK in the early 1990s, there was only one computer shared across his entire department:

> Whenever we needed to print something, we needed to copy the file to a floppy disk. Do you know the floppy disk? ... Copy to the floppy disk and then walk to the printing room and load it into the computer and print it. To me, that's already very, very nice. Then, by the time I went to the USA [in the mid-1990s], whoa, everybody has their own computer!

Hongyu was convinced that this lag in the availability and quality of research equipment and infrastructure continued to exist between the USA and the UK. Other returned Chinese scientists noted that regional differences in the quality of technological resources and research facilities no longer fell neatly along East–West lines. They argued that the research infrastructures in top Chinese universities was now on par with, if not superior to, what was available in Europe. One Chinese researcher who had trained in Europe noted with a laugh that, when Chinese scientists used to visit his European research facilities in the past, they would be impressed but that this was no longer the case as the research facilities in China improved:

> Before, when Chinese came here in Europe, ... they were always like, "Wow!" They take their cameras [out.] But now, if I bring them all, they will say, "Oh yeah, okay. My office may be even better."

Here, again, we have an indication of the power of the state to build up a country's research infrastructure through heavy investment, allowing scientists within the country to access the equipment and supplies they need to conduct high-quality research. This improved research infrastructure thus becomes another draw to recruit more overseas researchers to the country and to stem the brain drain of native-born scientists.

Conclusion

This chapter has recounted the many changes to the scientific research systems that returnees encountered when they moved back to Asia. In China, India, Singapore and Taiwan, returnees observed that significant improvements had taken place – from an increase in funding levels to the growing quality and availability of research facilities and equipment. But returnees also experienced negative aspects of their scientific context of return – in particular, inefficient and ineffective administrative processes that slowed down their research, and a sense of being on the periphery or semiperiphery of global research networks that were still largely centered in the West.

To address the latter issue, returnees were actively working to build new national and regional networks with more geographically proximate peer researchers. They were also trying to bring their West-based network contacts to Asia through invitations to attend conferences and seminars they organized. These efforts were paying dividends but there was not yet a sense that there was equality of network access for Asia-based researchers. The challenges with slow or ineffective administrative structures were harder to address, as they inevitably required more willingness from senior leaders within an organization to invest in an overhaul of such processes and also hire administrative support staff with the right research-oriented skillset and attitude.

Despite these hurdles, I found most of my returnees to be somewhat forgiving of the research systems in their particular Asian country, even as they pushed for improvements. Returnees expressed amazement at how the institutions they were joining in Asia had substantially raised their research support in the years since they had last been in the country. Karthik (whose story started this chapter) shared how much of a change there has been within Indian government circles in the last ten years in terms of desiring to make a significant impact on the global scientific stage. To him, the necessary building blocks were being put in place to make those ambitions realizable:

> Earlier, things were different. If anything was started, it was still under too much government regulation. You knew that [these institutions] were not going anywhere or they are [only] going to make a small dent in the kind of science that is done in India. On a global stage, they are not going to be prominent players at any point. That was quite clear. But by 2009, it was fairly certain that at least these institutions and potentially all the other IISERs that were coming up, … were going to lead Indian science into something that's globally prominent.

Still, there was a sense that while exciting research was happening in India, the scale at which this was occurring was not yet sufficient to make a significant difference in the country's global standing. There was also an acknowledgement that the sorry state of research administration structures in India resulted in significant wasted time on the part of scientists, holding back their research productivity.

Surprisingly, returnees in Singapore were also somewhat negative about the scientific research systems in that island-nation. I was not expecting to hear so many complaints given that the funding volumes in Singapore are very generous overall and the technical infrastructure top-notch. Several explanations are possible for this discrepancy. One is that the higher proportion of nonlocal scientists working in Singapore might make them less willing to resign themselves to the various research hurdles they encounter in a country they were not born in. Without a sense of national duty driving their decision to locate themselves in Singapore, they may be more exacting in their expectations of Singapore's scientific research systems. But it could also be the case that because Singapore positions itself as a "First World" country with the infrastructure to match, scientists who move there may be expecting a research environment that exactly mirrors what they experienced in the West, only to be disappointed when they encounter any shortfalls.

What I took away from these conversations was that no Asian country was without challenges as it attempted to upgrade its scientific research systems in the twenty-first century. While India was patently the furthest behind when it came to its research systems, it was also the place where the young ecologists I spoke with were the most excited about the networks they were building and the impact that they were having on the next generation of Indian ecologists. In Chapter 9, the concluding chapter, I offer some policy recommendations to India and my other case countries based on these findings. But first, the next chapter explores the scientific cultures in the research organizations where returnees worked and how they were trying to change specific aspects of these cultures to further improve their research productivity.

8 | Shifting Scientific Cultures in a Changing Asia

> There was this fantastic female student I had. A Singaporean. Very, very good. She came for a final-year project [with me] and I was trying to encourage her to stay on and do research [with me]. And she said that she would like to, but that the peer pressure is to go and get a bloody BMW! So she has gone into the world of marketing.
>
> India has so many people that it will be a long time [to see this change] but it has changed there also, perceptibly. The thing that you had for science – all that is slowly giving way to, "Do you work for a bank?" and "How much money do you make?" [But] it's a lot more acute in Singapore than in India. And that spills into the lab culture.

The quote above comes from a conversation I had with an Indian scientist, Yadav, who trained in the UK and the USA before moving to Singapore to join a research institute. Yadav was complaining about the difficulties he faced when trying to motivate the young people who worked for short research stints in his lab to choose science as a career over other more lucrative (and potentially easier) professional possibilities. This chapter deals with this different set of challenges faced by returning scientists. While Chapter 7 focused on the structural factors – funding availability, research equipment access, administrative processes – that impacted returned Asian scientists' productivity, this chapter discusses cultural factors in scientists' labs and classrooms (and in their research organizations) that also influenced their research experiences. Inevitably, these two sides of the scientific research environment (see Figure 1.4) influence each other. I focus more on the scientific cultures that returned scientists encountered in their Asian labs and classrooms, and the ways in which they were attempting to reshape different dimensions of these cultures to enhance their overall research experience.

Many of the returned scientists I talked with spoke at length about how they encouraged their students to engage with science differently from what these students were used to. In scientists' labs, this meant encouraging their students to treat scientific research with the same reverence that these scientists did, or to be more ambitious in their scientific research goals. In scientists' classrooms, this meant pushing their students to not only be more critical in their engagement with scientific knowledge, but also more curious and broadminded in their general approach to learning and problem-solving. These were norms and values that returnees felt that they had productively learned during their years of training in the West, and they were striving to inculcate the same values in their own students now that they were back in Asia. In other words, these efforts at cultural diffusion were a deliberate attempt by Asian scientists to reproduce elements of the scientific cultures that they encountered during their training in the West.

Before I discuss these efforts by returned Asian scientists, however, I first introduce the concept of "scientific remittances," which I coined to describe the social and cultural diffusion that occurs from destination to origin alongside the brain circulations and return migrations of scientists. I then discuss the various demographic, social and cultural changes that had been occurring in my Asian case countries while interviewees were overseas. These changes inevitably affected the education and scientific landscape in the four Asian case countries, such that interviewees effectively returned to a different country than what they had originally left behind. Only after going through these high-level changes, do I outline the four areas where returnees were focusing their energies to make changes to the scientific cultures of their labs and classrooms. These were: (1) encouraging a curiosity-driven approach to scientific learning, (2) raising their students' research ambitions, (3) trying to level attitudes towards rank within their labs, and finally, (4) broadening attitudes towards difference along various dimensions. These efforts were still a work-in-progress when I interviewed returnees, but in describing these change attempts, I am pointing to ways in which these returned scientists are trying to effect ground-up change in their respective Asian country's scientific landscape that could have a long-lasting impact on how science is conducted in these countries.

Scientific Remittances

It has long been understood that migrants are not just bodies moving through space and time, but that they carry with them specific knowledge, beliefs, values and ways of behaving and engaging with the world.[1] This cultural "baggage" is brought with them from their place of origin to their destination where it affects, and is in turn affected by, the culture and society of their new home. I explored one facet of this cultural assimilation in Chapter 4 when I discussed how interviewees were introduced to new ideas about the social practice and teaching of science when they first arrived in their Western country of training. The question that the present chapter explores is the process of cultural diffusion that occurs when some of these Western-trained Asian scientists return to Asia.

It is now well accepted that migrants continue to socially influence their origin communities even when they no longer live there.[2] In the late 1990s, sociologist Peggy Levitt coined the term "social remittances" to describe these "ideas, behaviors, identities, and social capital that flow from receiving- to sending-country communities" (1998:927). In Levitt's formulation, social remittances apply to immigrants who have settled overseas but maintain transnational ties with, and regularly visit, friends and family in the home community. The concept of social remittances is useful for explaining the transnational impact that these immigrants may have on their sending societies through cultural diffusion.[3] It complements the more well-known idea of "financial remittances," which represents the economic capital that immigrants send back home to enable the upward socioeconomic mobility of their families and communities in the sending country.[4]

In this chapter, I propose an expansion of our existing understanding of social remittances so that it also incorporates the ideas, values and norms that *returnees* bring with them when they move back home. In the context of Asian scientists' brain circulation, I further delineate a subcategory of social remittances that I call "scientific remittances." I define scientific remittances as the informational, reputational and

---

[1] Rapoport, Sardoschau and Silve 2020; Kulikoff 1986; Levitt 1998.
[2] Eckstein and Najam 2013.
[3] Levitt 1998; Levitt and Lamba-Nieves 2011.
[4] Katz and Stark 1986; Stark and Bloom 1985.

cultural diffusion that occurs as a result of the brain circulation of scientists:[5]

(1) Informational diffusion refers to the scientific knowledge transfer that occurs when overseas-trained scientists bring their increased scientific knowledge and research experience back to research organizations in the home country,
(2) Reputational diffusion refers to the greater name recognition, network access and network standing that a research organization in the home country enjoys when a well-connected and well-respected scientist joins their faculty ranks from overseas, and
(3) Cultural diffusion refers to the new social and cultural norms and values related to the practice of scientific research and teaching of scientific knowledge that are brought back by returning scientists.

In the past, Asian governments encouraged the Western training of their future scientists in the hopes that they would benefit from such scientific remittances which can begin, even without any return migration, through the establishment of transnational research partnerships, participation in international conferences back home, as well as through transnational communications with students from the home country who may reach out to the overseas scientist.[6] Vanya, who returned to work in a research institute in India, shared how she knew of many Indian scientists who had opted to stay in the West but who still found ways to stay connected and contribute to the scientific community in India:

> One way they stayed connected was to at least come back every year or every other year. They've given talks; they have actively engaged with the Indian system of science. They haven't forgotten their original roots and they have contributed in many ways. I can think of multiple people who have been on advisory committees and helped with overseeing things. You know, I can think of so many people at [my research institute] who came [during their] sabbatical and then either taught courses or contributed in different ways. It's very valuable because they bring with them a dimension that a lot of our students don't see. And when we have somebody like that come and give a talk and do

---

[5] Ackers 2005.
[6] Le 2008; Robertson 2006.

a three-month course and actually teach, it can be immensely valuable to those who chose not to go [out of the country]. They can really get enriched by having that experience.

As valuable as this form of scientific remittance via visiting scientists can be, cultural diffusion processes get kicked into high gear when Western-trained Asian scientists *return permanently* to teach, set up their own labs in Asian research organizations and work side-by-side with young scientists-in-training. The scientific remittances that returning Asian scientists bring back with them include not just new scientific know-how and network connections in the global scientific field, but also new norms and values regarding the social practice of scientific research. Their long exposure to scientific cultures in various Western countries shapes their preferences and priorities when they put together their own research team, teach classes, mentor young researchers-in-training and pursue their particular research agenda back in Asia.

It is clear that Asian governments want returning scientists to bring change with them. But these governments may prioritize the informational and reputational diffusion that occurs as a result of return, forgetting that cultural diffusions around scientific norms and values are also occurring.[7] In many cases, Asian governments established new research institutes designed after the administrative structures, hiring and budgeting policies and general research environment found in Western, and specifically American, universities and research institutes, rather than trying to reform from within. However I argue that the ground-up cultural changes occurring in Asian labs, university departments and research institutes are just as important as these large-scale structural investments that select Asian governments are making to improve the overall scientific environment in their country. As one of my interviewees in Singapore shared with me: "You have to develop the culture. You cannot just throw money. You can hire maybe the biggest name, but unless you change the culture, I don't think it's going to last."

---

[7] Programs like the Chinese government's Thousand Talents Program, which allows highly acclaimed foreign faculty to spend a few months each year in China, is an example of a program that appears to be largely focused on helping Chinese universities accrue the reputational benefits of having a superstar scientist affiliated with their organization, rather than trying to effect deeper institutional change.

It is for these reasons that this chapter and the previous one should be read in tandem. While Chapter 7 explored the changes that governments have made and are making to scientific research systems in my four Asian case countries, this chapter focuses on the changes individual returnees are pushing for in the scientific cultures of their research teams and science classrooms in Asia. While research system changes were largely initiated in a top-down manner by the state or by the leadership of research universities and institutes, changes to scientific cultures are being introduced by returning scientists themselves at the level of one-on-one interactions. At the same time, these cultural changes that returning scientists are trying to instigate are interacting with broader societal shifts in these Asian countries that are a result of their rising income levels, growing middle class and evolving economic base, I discuss this next.

A Changing Asia

When studying returning scientists' experiences trying to reshape scientific cultures back in Asia, it is necessary to acknowledge how both the country they returned to and the returnee had changed during their years "apart" and to look at how these twin change processes interact with one other. The subtitle of this book – *Changing Science in a Changing Asia* – is a reminder to readers that my four Asian case countries were undergoing rapid societal changes before and during the years that my interviewees were training overseas. These Asian societies were not empty vessels into which returning migrants poured their own cultural values and norms; nor had these societies' cultures remained static while interviewees were overseas. Keeping in mind these broader societal and intergenerational changes highlights how a process of dissimilation, or a growing difference, between the migrant and their origin community can occur. Even if the migrant does not change, their origin society may still undergo change while they are gone.[8]

There are several demographic, social and economic changes that my four Asian countries underwent over the last few decades. I have already written in Chapter 2 about the growing investments in science and technology in all four countries. What I have not discussed is how,

[8] Fitzgerald 2013.

in some of these countries in recent years, there has been declining interest in careers in the natural sciences among the young. Another societal change we are seeing in all high-income Asian countries is a demographic shift as birth rates rapidly decline, leading to fewer native-born young people available for graduate education. I elaborate on these societal shifts in this section because these changes were also felt in the labs and classrooms of returning Asian scientists.

In Taiwan, there has been a marked shift away from the pursuit of doctoral degrees. As education scholars Chen-Wei Chang and Wang-Chin Shaw noted in 2016, "a doctoral degree used to be an honor in Taiwanese society; today it has become a burden." These authors note that the number of doctoral students grew rapidly between 1984 (when there were only 1,500) and 2014 (when there were 30,549). The peak number of doctoral students in Taiwan occurred in 2011 when there 34,178. Since 2011, however, there has been an almost year-on-year decline in the number of students pursuing a doctoral degree.[9] By 2019, there were only 28,510 doctoral students in Taiwan, and only 1,530 of them were pursuing PhDs in the life sciences.[10] Some of this stems from Taiwan's declining population, which has caused many of its universities and higher education institutions to merge or shut down in recent years.[11] But this shift away from doctoral studies also has to do with the limited job opportunities available for PhD holders in Taiwan.[12] Yet another factor behind the declining number of doctoral students is the fact that Taiwanese students are shifting to other careers that do not require as much advanced training. In 2019, there were roughly eight times as many tertiary students pursuing degrees in business administration as compared to students pursuing degrees in the life sciences in Taiwan.[13]

The same can be said for Singapore – another small, high-income country with a declining birth rate. The number of graduate degrees in business administration (2,524) granted by all of Singapore's

---

[9] CEIC Data, accessed on July 10, 2021, www.ceicdata.com.
[10] Directorate-General of Budget, Accounting and Statistics, "Statistical Yearbook of the Republic of China, 2019" (Directorate-General of Budget, Accounting and Statistics, 2020), 30, table 15, accessed on March 11, 2021, https://eng.dgbas.gov.tw/public/data/dgbas03/bs2/yearbook_eng/Yearbook2019.pdf.
[11] Chou 2014.
[12] Chang and Shaw 2016.
[13] Directorate-General of Budget, Accounting and Statistics, "Statistical Yearbook," accessed on March 6, 2021, 30, table 15.

A Changing Asia   265

autonomous universities in 2019 was approximately four times the number of graduate degrees in the natural, physical and mathematical sciences (676).[14] While the sciences used to be one of the most popular choices in the country, especially for female students' first university degree in the 1990s, by 2019, it had dropped considerably in popularity across both male and female students.[15]

Unlike Taiwan, however, Singapore has been much more successful at encouraging immigration into the country to bolster its falling native birth rate. It has also been more successful at internationalizing its higher education sector. As mentioned in Chapter 2, there is a higher proportion of international students in Singapore's top tertiary institutions as compared with Taiwan. In addition, the international student population in Singapore is more diverse, coming from various countries in East, Southeast and South Asia. Taiwan, in contrast, largely draws its international students from East Asia (mainly Macau, Hong Kong and mainland China). This is partly because of language policy differences between the two countries, given that the language of instruction in Singapore is English while it is Mandarin Chinese in Taiwan, but it is also because of a widespread sense that Singapore is a more inclusive society than Taiwan.

In large countries like China and India, a slight decline in undergraduate student interest in the sciences will not have as much of an impact at the graduate level. The sheer volume of people in both countries means that there will still be many students committed to continuing their studies in the sciences. However, China does have a demographic crisis looming on the horizon as its college-age population is expected to shrink by roughly 40 percent between 2010 and 2025.[16] In contrast, the number of students in India enrolling in master of science programs between 2014 and 2018 continued to increase each year, with more than 600,000 students enrolling in such master's

[14] "M850551 – Graduates from Higher Degree Courses by Type of Course, Annual," Department of Statistics, Singapore, updated on January 21, 2021, accessed on March 11, 2021, www.tablebuilder.singstat.gov.sg/publicfacing/createDataTable.action?refId=15200.
[15] "M850511 – Graduates from University First Degree Courses by Type of Course and Sex, Annual," Department of Statistics, Singapore, updated on January 21, 2021, accessed on March 11, 2021, www.tablebuilder.singstat.gov.sg/publicfacing/createDataTable.action?refId=15207.
[16] Gu, Michael and Zheng, "Education in China."

programs in 2018–2019.[17] The proportion of women in these master's programs in India has also steadily increased, with 174 women enrolling for every 100 men in 2018–2019.[18] Out of my four case countries, India is best positioned demographically to increase its scientific workforce and output in the coming decades. However, as noted in Chapter 2, India's ongoing challenge is its fractured and uneven higher education system, which has yet to be addressed by the central government. As one of my interviewees who returned to India put it:

> The [Indian] education system – ... they don't teach you to think, they don't teach you to write. It's all just learning information. They don't teach you to analyze anything so if you're not keen enough to have a mind that works that way and family with a lot [of] intelligence, then obviously you're not going to do better.
>
> But we are a billion people, you know? ... Even if only 0.01% of the population is bright, then those people are going to stand out. And that's still very large in terms of total numbers. But the education [system] is very bad. Very uneven.

In contrast, China has undertaken large-scale reforms in its educational system in recent years, partly because it has accurately forecasted the future dangers that will come with its aging population. In 2001, a New Curriculum Reform was kickstarted to shift its education approach from a narrow, specialized knowledge transfer process focused on specific subjects, to a broader and more internally integrated one.[19] Still, as with India, there continue to be wide discrepancies in the quality of education available to children in urban versus rural parts of China, and this shows in their differential acceptance rates into the country's top universities. Between 2009 and 2014, 97 percent of China's poorest counties sent no students to Tsinghua University, China's top university.[20] Children of rural migrants who follow their parents to urban centers also face difficulties when it comes to higher education opportunities, as they lack local registration (or *hukou*).[21] Without addressing these inequities in educational access

[17] Department of Higher Education 2019, figure 59.
[18] Department of Higher Education 2019, box 20.
[19] OECD, *Education in China: A Snapshot* (OECD, 2016), accessed on March 9, 2021, www.oecd.org/education/Education-in-China-a-snapshot.pdf.
[20] OECD, *Education in China*.
[21] OECD, *Education in China*.

and quality across the country, China will also not be able to reap the full potential of its large population as it seeks to increase its scientific output.

The recent and upcoming societal developments listed here serve as a backdrop to the increasing investments in science and technology by my four Asian case countries. It was to this evolving landscape that Asian scientists returned, armed with their own ideas about the additional changes they wanted to introduce as well.

Returnees as Change Agents

In their efforts trying to effect cultural change in their new workplaces in Asia, all returnees reported encountering challenges. This is to be expected, not just because Asian societies had undergone change while returnees were away but also because cultural change can be difficult to direct or coordinate. Writing about Indian software professionals who moved to Bangalore after years of working in American software companies, anthropologist Carol Upadhya notes that the highly skilled returnees she interviewed experienced similar difficulties when they tried to effect change. They came back to Bangalore with "new ideas about modernity and proper civic life through which they reenvision[ed] the city as a global place" (2013:142). But their ambitions ran into problems when these migrants encountered an inertia in their home country context that resisted cultural change. Political scientist Jean-Pierre Cassarino also notes that even though some returning migrants might engage in a "return of innovation," believing that they can bring new ways of doing things to their native countries, they may run up against structural resistance to their efforts, which could ultimately stymie them.[22]

Perhaps in response to these challenges, but also because most of my interviewees returned at a relatively junior stage in their careers, they tended to focus their efforts more narrowly on changing the scientific cultures in their labs and classrooms. Within the broader scientific environment in their particular Asian country or organization, they saw the domain of their lab or classroom as a space that was more directly within their locus of influence. And so they concentrated their efforts on (re)shaping the scientific worldviews and priorities of their students and

[22] Cassarino 2004.

research staff. It was clear in how interviewees spoke that they did not see themselves as "just" researchers in pursuit of new scientific knowledge and discoveries. Instead, returnees saw themselves as mentors to the next generation of Asian scientists. Vidita, a returning scientist in India, shared how she found this role both difficult and exhilarating:

> Sometimes, it can be frustrating. But it can also be very exciting if you treat it as an opportunity to do something which is quite different from what most scientists have to do. Which is really build, in the sense of building the base of your own environment, getting the opportunity to mentor young people.

Likewise, Li Jie, a Chinese scientist who had chosen, upon leaving the USA, to work in Singapore rather than China, saw a key part of his professional mission as improving the state of science education in Singapore:

> There are two things I want to do. The first thing is developing technology, you know, to benefit the people. Second thing I want to do is have a high impact in terms of educating more young people. Because I just feel, if you are working alone, if you are only one, it's so hard to make a high impact in the society. You have to educate a lot of people.

It was irrelevant to Li Jie that he was educating mainly Singaporean, and not Chinese, students. From his point of view, cultivating the appropriate norms and values of scientific inquiry in his students was simply part of his responsibility as a scientist, irrespective of where he was in Asia.

At the same time, these returnees were also interested in transforming the scientific culture in their new institutions for more pragmatic reasons. A lab culture that encouraged long hours, a single-minded dedication to research, and creative problem-solving in the pursuit of scientific knowledge helped these scientists as the PIs of these labs, in the same way that these scientists had helped their own PIs when they were trainees earlier in their careers.

Drawing on the framework I first introduced in Chapter 4, I use the present chapter to elaborate on a select few dimensions of the scientific cultures that returnees were trying to change in their new Asian classrooms and labs. While Chapter 7 indicated the structural changes returning scientists wanted in scientific research systems in their Asian workplace – such as improving the research administrative functions in their new place of work, and building more research network connections within Asia – here I focus on four key cultural dimensions where returnees identified a need for change:

(1) Encouraging a curiosity-driven approach to scientific learning,
(2) Raising their students' research passions and ambitions,
(3) Leveling attitudes towards rank within their labs, and
(4) Broadening attitudes towards difference.

In the following sections, I discuss both the changes scientists were trying to enact as well as the particular roadblocks they encountered along the way.

Encouraging Curiosity-Driven Scientific Learning

Even as they acknowledged significant improvements in the availability and quality of research-related technologies and services in their particular Asian country, returnees were concerned that the contemporary approach to science instruction in Asian educational institutions was still too oriented towards rote learning. They worried that this approach killed most students' interest in scientific exploration and also in science as a career.

Returnees were also adamant about the need to revamp the teaching of science at all levels in their respective Asian country so as to encourage a broader interest in multiple scientific fields, rather than a narrow, siloed focus on a single subfield. They saw this educational reform as a way to stoke students' general curiosity about the natural world and thereby set the stage for more cross-cultivation of ideas drawn from different scientific fields. Nihil, an ecologist at a research institute in India, lamented the lack of an open-minded interest in new ideas and new topics that he encountered among the graduate students at his institute:

> We have this seminar series and I am trying to get faculty and students to attend the seminar series. But unless we have Madhav Gadgil [who is a famous Indian ecologist], we can't fill the room. It's that culture. In [my US doctoral university], it wasn't an Ivy League school but you could have filled that hall. Tuesday at four o'clock, everybody was there, and you had to fight to find a chair, and there were hundreds of people. It was exciting to be part of that. ... We should spread that culture here.

Indian researchers like Nihil complained that India's moribund educational system was not stimulating Indian youths' scientific imaginations. They were hopeful about the long-term impact of the new Indian Institutes of Science Education and Research (IISERs) that had been set up around the country, but noted that there were many older universities throughout India which were not offering an innovative science curriculum or encouraging a general inquisitiveness about the natural world among their students. According to Nihil, without a wider transformation of the science education system in India, the vast majority of Indian students would not benefit from the pedagogical innovations occurring in the IISERs. Kannan, another Indian returnee, spoke about how he needed to reeducate his students to be more open to new subfields after they joined his research institute, and how difficult this could be:

> Very few of them are adventurous. Very few of them have a broader perspective of what science is and what kind of science different people are doing. And even after they come to [my research institute], they really don't change much.

Returnees in my other three Asian case countries also shared that higher educational institutions in their particular country did not yet enjoy a robust culture celebrating the open exploration of new ideas from disparate fields. Like Nihil earlier, Chinese scientist Eng Chye, who relocated to Singapore after spending many years in Europe, shared how it was the norm among students and faculty at his European doctoral and postdoctoral universities to buy a simple take-away lunch and then attend public research talks organized by different departments during their lunch hour. Even if the topic was not related to their area of specialization, everyone still attended so that they could learn something new. Eng Chye complained that in Singapore, in contrast, he had to constantly nudge his students to attend research talks during their free time instead of spending that time surfing the web: "I already told my students in Singapore, 'Hey, why were you going on the internet? It's okay if you just go and listen. Research is really knowledge.'"

Jian Kai, a Taiwanese scientist who had chosen not to leave the USA, was adamant that early exposure to a variety of fields was essential if one wanted to be successful in biotechnology, given how applied this area of innovation was. Reflecting on his own success in biotechnology, he surmised that it was his initial training in physics that had enabled him to do well in this field:

> Biotech right now is a combination of physics, engineering, chemistry and biology. So anybody who just studied biology cannot be good. Anybody who just studied physics cannot be good here.
>
> You know, I spent two years in physics – just by accident. And later, I also studied chemistry, biology, all this. And at that time, I thought I wasted my time, but now I feel – actually no – those are all my skills. Because I can communicate with engineers, physicists, chemists very easily, just because of my previous education.

According to Jian Kai, Taiwanese students in high school and beyond needed to be exposed to a broader variety of fields to ease their communication with people from a range of backgrounds, and to think in a more interdisciplinary manner. As with many other interviewees, Jian Kai felt that the USA excelled in this approach compared to Asian countries.

Returnees acknowledged that without a change in the primary and secondary educational system in their particular Asian country, their attempts at changing students' attitudes towards scientific learning later in life would remain an uphill battle, given that they only came into contact with these students at the tertiary level or even later. However, despite these difficulties encountered when (re)training their students, returnees still thought it was necessary, especially as they hoped to convince at least some of these students to pursue graduate studies and careers in science.

Raising Research Passions and Ambitions

Returnees also actively encouraged their research students to be more ambitious when setting their research goals and to be more passionate about their research in general. This was often a challenge as their students did not always share the same outlook about science as returnees did.

Stimulating a Passion for Research

Li Jie, a Chinese scientist working in a Singapore university, shared how he tried to motivate the undergraduates in his class to treat their lab work as more than just a class assignment:

I have twenty undergraduates in my group. I'm trying to motivate them because Singapore is really a place where people are not interested in science. They just ask, "What's useful? Where will this be used?" I try to motivate them and say, "You know, if we do this and make this, then maybe we can make a patent, build a company, you know." But it seems pretty hard for me to convince people here. They're just not interested. Especially for local people, they don't get the point. [They think,] "I got a degree, so I can get a job easily." I talk with my undergraduates and I try to convince them to do a PhD, but their answer is: "If I do a PhD, I spend four years here, every month I get maybe SGD 2,500 per month. And after four years, I am still looking for a job. But if I work in a company right now, I make SGD 3,000 per month. And after four years, my salary will go up."

This attitude was different from what interviewees had been exposed to when they had trained in the West. During their graduate training in the West, interviewees were caught up in a general passion for scientific research that was routinely shared by their fellow trainees, creating a self-feeding mechanism that pushed all of them to work harder. Like my scientist interviewees, many of their fellow trainees had also been international students. Vinu, who moved to the USA in the 1990s for his PhD, joked about how it was always the Indian and Chinese graduate students, and not the American students, who worked the longest hours at his East Coast university:

At [my US university,] we used to have a last shuttle [bus] at 6:30 p.m. in the evening and we would call it the "Indo-China Shuttle," because on a Saturday evening at 6:30 p.m., you are unlikely to see any Americans getting out from the lab! [*Vinu laughs*] That's why the Indian and Chinese students do so well.

In contrast, returnees in Singapore and Taiwan noticed a distinct lack of such passion for science among many of their students in these two Asian countries. Li Jie felt that the graduate students he encountered in the USA had been more committed to research and willing to put in much longer hours to produce results, unlike the graduate students he encountered in Singapore:

Here [in Singapore], it's just too slow, you know? My students come in at 9:00 [a.m.] and leave at 6:00 [p.m.]. They come to work Monday to Friday. They take this as a job.... But if you go to the United States, you see often people working in the night, on the weekend. In the US, I only take, maybe, one vacation per year. Like two weeks is maximum. But here, my students always use out all their energy in the weekend and holidays. They enjoy every day!

Of course, some of these complaints from my interviewees could be attributed to the stereotypical intergenerational griping that paints younger people as being less motivated, lazier and not as resilient as the older generation before them. But interviewees were also able to draw distinctions between students from different Asian countries, which indicates that this was not simply an intergenerational issue. Yadav, the Indian scientist working in Singapore who was introduced at the start of this chapter, contrasted the attitude of present-day undergraduate students in India versus in Singapore:

> There is not a chance that, after six o'clock, I'll see a student in my lab [in Singapore]. Whereas in [the top institute in India], you know, students are running after [the professor] at midnight. So that hunger, the culture, still remains [in India]. [A colleague in India] has his lab meetings on a Sunday, you know? I was quite amazed! Here [in Singapore,] not a soul will turn up. They don't turn up after 6 p.m. here.

This ongoing enthusiasm for the sciences that Yadav observed in top Indian universities relates to the point I raised earlier in this chapter, about how India's large population means that even if only a small fraction of students in the country are passionate about science, this will still translate into a large number of enthusiastic science students.

Likewise, Li Jie shared that he had much more success with non-Singaporean students in his lab. He had managed to recruit a very international team with graduate students from China, India, Indonesia and Iran, and he found that, on average, they were more committed than his local students when it came to their lab work, paralleling the drive of foreign students in the USA:[23]

---

[23] Such student research teams comprising multiple Asian nationalities speaks to the increasing intra-Asian mobility occurring at the training moment, with more Asian students considering migrating to another Asian country (like Singapore) that can serve as a regional training hub and sometimes also as a stepping-stone to a destination in the West at a later point in the student's career course.

Of course, it's a bit easier to convince foreigners. . . . I feel they're more motivated. They have more energy to do work. I guess maybe that's also the reason why America is so powerful in terms of science. Because they hire immigrants over there. The only way these immigrants can be successful is to work hard to do something great. That's the only way you can become successful. But for local people, well, it doesn't matter what they do, right? They will find a way to be successful.

For these reasons, Lie Jie and other returnees I spoke with in Singapore and Taiwan wanted to increase the international makeup of their lab teams, seeing diversity along nationality lines as a way to inject new perspectives and also new energy into their research team. But, as I discuss later, diversifying the pool of graduate students and postdoctoral trainees in their Asian labs was often hard to accomplish.

Increasing Research Ambitions

Having come from labs in the West where their supervisors were often leaders in their field, returnees had learned to dream big themselves. Upon their return to Asia, they wanted to inculcate similarly large research ambitions in their students. But instead, returnees remarked on how the students they encountered back in Asia did not seem as interested in solving big, complex problems, but tended to aim their research goals lower. Andrew, a Taiwanese clinician-scientist who taught at a top university in the USA before returning to Taiwan, shared how he tried to encourage his local graduate students to stretch themselves by choosing challenging and ambitious research problems to solve:

> "Pick a very, very difficult problem that needs to be solved. And spend your whole life trying to solve it." That is what graduate school should be. But I don't really see that in Taiwan yet. Students, maybe because they're Asians, they are more reserved, they are less ambitious. They think, "Okay, this is too big a problem. Big problems are for someone at Oxford or Cambridge to solve." And I don't like that. I would like students at least give it a try. Of course, we may not succeed but the whole point of being in a graduate program is not to find the easy way out. . . . But, I don't sense that ambition kicking in, you know? . . . If the society is going to move forward, if we want to

improve the well-being of mankind, someone needs to take on the challenge of these very difficult problems.

Andrew's theory that his students' Asian cultural background made them less ambitious does not jive with the grand aspirations of the governments of my four Asian case countries. Returnees in all four countries noted that their governments had grand plans for expanding the volume and impact of the scientific research coming from their respective countries. Returnees like Andrew saw part of their job scope as encouraging their student research assistants to also aim high and take advantage of the many research opportunities now available in Asia. But other returnees had alternative explanations as to why their students were not as ambitious as they would have liked.

Hualing, a Chinese researcher who spent many years in Europe before returning to China, hypothesized that the easy availability of research funds in present-day China paradoxically led new trainees in that country to work on easy questions rather than challenge themselves with ambitious, boundary-pushing questions:

> If you're ready, you can do good science [in China]. Really, you don't have to worry about the funding. So that's good. But ... the students – they do not like to challenge. So that's a problem because, in research, you have to be creative, right? You have to do something really new. You have to challenge the existing theory and knowledge. That's the soul of research. But back in China, unfortunately, it's not. They work hard, but they don't push the boundary. So when you look at the number of publications, China is already the number one, I think. But in terms of the high quality of the papers, it's not. And that's the problem.

Meanwhile, a returnee in Singapore argued that the performance metrics in his country were too geared towards short-term, quantitative measures, such as a scientist's annual number of publications and their citation counts, and that this reward structure led researchers to focus on relatively easier research questions where positive (but also less path-breaking) results were guaranteed:

> At this moment, the [Singapore] community is divided over what constitutes "good science." I don't agree with how people measure science here. Here, it's about the number of papers, the journal impact-factor, etc. To me, that is not good science.

> Right now, the culture is definitely one of "what do I need to do to get my paper?", "What do I need to do to get my money and tell me what to do," rather than "I am doing this for its own sake." And so, being internally driven versus externally driven. And that creates a very different kind of science.

But Madhu, a scientist in India, had yet another explanation for his Indian students' lack of ambition. To him, it stemmed from their lack of self-confidence. He shared with me that while the students in his lab were well trained technically and innately creative, they had never been encouraged during their earlier school years to think of themselves as being able to solve the "big" problems in science. Because of this, Madhu mentored his students extensively to help them internalize a sense of themselves as confident and capable scientists, on par with the best in the world:

> The big difference is that self-confidence is very low for many of the students here. So initially, there's a lot of hand-holding and prodding to build them up. But that also eats up a lot of [my] time. So the number of students I think I can simultaneously mentor here is far fewer than what I can.

Madhu's efforts are commendable even as he acknowledged that this meant he was only able to mentor a small number of students in his lab each year. His point echoes the argument made by Andrew at the start of this section that his Taiwanese students thought that "big problems are for someone at Oxford or Cambridge to solve." Expanding aspiring Asian scientists' belief in their own capabilities, and their sense of what they could achieve in Asia itself without having to relocate to the West, was perhaps the hardest challenge confronting returnees. And it is one that also requires a decolonization of the educational systems in Asian school and universities, and different reward and incentive structures to be built into their research funding systems.

Leveling Attitudes towards Rank

A key characteristic of many of the labs and classrooms where returnees had trained while in the West was the egalitarian environment they possessed, where free debate between faculty and students was the norm. Returnees saw this attitude as linked to a firm belief in the meritocracy of ideas (along the same lines as Robert Merton's principle of universalism[24]), and they sought to nurture a similarly flat hierarchy within their labs and also within their research organization. This was not always easy to implement, as many of their students came from backgrounds where it was normative to show respect to one's elders and seniors by agreeing with them or, at the very least, not disagreeing with them. Yuewang, a Chinese scientist in Singapore, spoke of how difficult it was for him to convince the staff and students working in his lab to critique his ideas even though he was their supervisor. He shared with me the challenges he faced while trying to stoke a debate over different ideas in his lab group:

> I think the students, or kids in general, are very much tamed, you know? The educational system here is a system where people are so willing to listen to you. I think that's toxic to science.... They are not willing to challenge and they are not willing to rebel. They are not willing to say, "I want to take a different view." ... I feel there is very little of that spirit. Maybe I am not correct, but I don't feel that spirit here so much. That sort of contesting and really enjoying the clash of different ideas. In most scientific settings, you see your limitations by seeing other people's strengths. But here, there is no such setting.

Meanwhile, Hoe Fang, who had returned to Taiwan, spoke of what he saw as the incompatibility of the Confucian norm of respect for one's elders and the scientific norm of questioning received wisdom. He had returned to Taiwan, knowing he would likely experience a reverse culture shock, but he had not thought that he would grow so frustrated with the scientific culture in his birth country. He placed the responsibility squarely on Taiwan's Confucian culture, which he described as "important for society but bad for science":

> Confucius told us to respect the elder's experience, so we have to listen to them. Even if they are incompetent, we still have to respect them because Confucius told us to ignore incompetence in the elder.... It's bad for doing science, but it's good for other aspects.... Such unique culture thing, it's not good for doing science here. Because in the United States, if you are getting older, you are expected to become an even greater scientist. If not, you will be fading out very soon, and the resource will be given to young people. But that's very different here.

---

[24] Merton 1973.

Even in India, a country without a Confucian culture, scientists talked about how the deference expected to be shown elders and those with higher rank stifled the free debate of ideas in scientific settings. Divya, a US-trained scientist who returned to work at a research institute in India, talked about how difficult it had initially been for her to inculcate a hierarchy-free culture in her lab because that was not how Indian students were trained to behave:

> The hierarchy is another thing that is vastly different, for sure. I know that in most universities in India the hierarchy is just established and is well accepted. And so, there's no effort to really get rid of it either. Even though, in my lab, I'd like there to not be a hierarchy. I think that the way that Indian students grow up is expecting it. Like it's just a given in India. You struggle with that a little bit. But eventually, once you have a lab culture that forms, then I think it's much easier for an incoming student to just conform to it. And so, you want to just make sure that your lab culture to start with is what you want it to be.

Like Divya, most returnees were focused on shaping their lab culture to replicate the flatter hierarchy they had enjoyed during their training in the West. This meant instituting regular lab meetings where everyone in the team (including themselves) had to present their latest work and receive feedback, encouraging the formation of journal clubs where students had to discuss and critique published work, and assigning presentation work in their classes with a follow-up question-and-answer session. All of this was to encourage the establishment of a culture in their labs and classrooms where it was normative to debate ideas regardless of where (or who) these ideas came from. These efforts were also aimed at changing the communication culture within their labs, by encouraging everyone to share their research progress with each other and invite questions.

Returnees noted how it took a long time to nurture such traits in their students and get them to the point where these students were comfortable with an open approach to discussion that could include disagreeing with their professors. But returnees expressed a willingness to make this time commitment to effect such attitudinal and behavioral changes in their students. Tsin Yee, who had returned to Singapore in the early 2000s, noted that the Asian doctoral students she accepted into her lab tended to start out rather quiet, but that, by the end of their training with her, she had mentored them into becoming more forthright in stating their opinions:

> The PhD students that we get here, they tend to be very quiet. They are like the workhorse in the lab. They don't talk up as much with the professor. But by the time they finish their PhD [with me], when they are more mature, they are able to speak up more.

As previously described in the examples of Madhu and Divya, returnees like Tsin Yee were investing their time to mentor the young Asian scientists-in-training in their labs to be more critical of received wisdom and more questioning of authority. These returnees were making small-scale changes in their lab and classroom cultures in the hopes that this would eventually scale up and lead to larger and longer-lasting cultural shifts in their research organizations. These efforts were bolstered by the fact that these Asian societies are manifesting more individualization (and in certain cases, also increasing individualism) over time, and so this broader societal change should make it easier to instill a flatter hierarchy in research labs and organizations in these Asian countries in the long run.[25]

Broadening Attitudes towards Difference

Another dimension of scientific cultures that many returnees tried to change in their Asian classrooms and labs related to institutional attitudes towards diversity. As mentioned previously in Chapter 4, returnees who had trained in the USA spoke about how much they had appreciated the diversity of staff and students at their American training institutions, and how they wanted to increase the diversity within their Asian organizations along similar lines. They saw a positive correlation between the diversity within an organization and the richness of research it generates. Most returnees reported that the scientific cultures in their new Asian workplaces were not as welcoming of foreign scientists as the United States had been, though they also agreed that European countries did not do well in this regard either. Anthony, a clinician scientist in Taiwan, insisted that the USA had it right and

---

[25] Hamamura 2011; Chang 2014.

talked about the importance of inclusivity in unlocking the full potential of research teams:

> One thing I learned at [my US university] is the importance of diversity. ... Your diversity will ensure the vibrancy of different views. ... So it's not just lip service but there is value in diversity, whether it's across gender, across ethnic background, across different countries. Because, you know, the twenty-first century is going to be a global village. So the earlier we recognize the diverse nature of human beings, I think that institution will work much better.

Within Asia, certain countries were deemed to be better at opening up their scientific research sector to foreigners. Li Wei, a senior researcher at a research institute in Taiwan, shared that he found Taiwan to be less open to foreigners as compared to Singapore, but still more inclusive than other East Asian countries like South Korea and Japan. Likewise, when asked how he would grade Taiwan with respect to its diversity, Andrew shared that both geopolitics and language differences kept his institution from being more inclusive of foreigners, whether they came from China or from outside East Asia:

> I would give Taiwan a D [grade]. I would say the general culture is actually warm and friendly, but it doesn't embrace diversity, alright? ... One reason is the politics. It's kind of a big animosity towards certain segments of China. Like, you know, Chinese students. Why can't more come to my university? I think like 10 percent of all graduate students in Hong Kong are from China. Why can't we do that? In my building, there are students from India, Malaysia and other countries, but still not a lot. And I think that's where our training is somewhat deficient. Also because of the language barrier. Many faculty are trained in the Western world but, by and large, the lectures and all that here are mostly in Chinese. And therefore it has been a barrier.

Returnees spoke of a range of tactics they had adopted in an effort to "internationalize" their lab teams. Interviewees in China told me how they would insist that their doctoral and postdoctoral trainees make their research presentations in English. Returnees in Taiwan shared that, if they had foreign trainees in their lab, they would require everyone in the lab to speak English at all times. However, these scientists also reported that it was difficult to enforce this rule and that nationality-based cliques would quickly form within their labs. Carol, a Taiwanese scientist, admitted that she had given up on recruiting foreign students into her lab because it caused interpersonal tensions within her lab team, which she then had to personally resolve. Despite a steadfast belief in the learning power of diverse labs, it was less of a headache for her to recruit only Taiwanese students and so she had resigned herself to this outcome.

Furthermore, even as they spoke of the importance of nationality-driven diversity amongst their trainees, not all returnees were as keen about internationalization at the faculty level. This was particularly true in China and Singapore which had aggressively recruited foreign faculty to join local universities and research institutes in a manner that alienated some native-born scientists. I bring up this point to show that even when the leadership of research organizations and returning scientists share a similar goal – increasing the diversity of these institutions – they could have divergent ideas about how to achieve this desired end state.

Singapore's decision to make English its national language of education and business has made it an attractive destination for foreign investment and partnerships. But Singapore's example is also illustrative of the challenges that Asian countries may face as they attempt to internationalize their faculty. The question of how to treat foreign faculty, and whether or not there is (or should be) an even playing field between locals and foreigners, and also between different categories of foreigners, was an issue that many of my interviewees in Singapore – foreign and local – raised. Some of the Singaporean scientists I interviewed who had returned to work in Singapore complained about what they saw as a biased system that privileged expatriate faculty over local faculty:

> The pay is very different. The benefits are different. Usually, the foreigners come in at a higher level than the locals, with a more attractive recruitment package. The thinking is that there is an opportunity cost to come here from Europe or from the US. But I don't think there is any such thing as an opportunity cost. You should be paying people according to their work and their productivity. It is not about Singaporeans versus foreigners. It's about doing the right thing. You shouldn't have to wait until Singapore is a more attractive destination before you do the right thing.

In this regard, Singaporean faculty were drawing a distinction between the approach taken by Singapore to foster diversity versus the approach in the United States. They felt that the latter country managed to combine diversity with meritocracy more successfully.[26] Even expatriate Asian faculty I interviewed in Singapore complained about how Singapore approached its diversity goals. A Chinese national, who worked in Singapore for several years before eventually returning to China, complained about what he saw as a two-tier expatriate system in Singapore that gave a more generous benefits package to Western expatriate faculty than to Asian expatriate faculty:

> You see everywhere, there are Europeans or Americans, but they look down on the Asians. This is the very bad thing in Singapore. ... Can I tell you a story? Our boss [in Singapore] tells us the Europeans are paid higher salary and Asians lower. He just says, "Europeans need to eat cheese and cheese is expensive." So then I say, "The Chinese usually eat bird's nest,[27] and isn't that more expensive than cheese?"

These tensions between locals and foreigners existed in China as well, as a result of the Chinese government's efforts to recruit foreign faculty through its Thousand Talents Program. Returning scientists in China complained that the terms offered to these foreign faculty were overly generous, allowing them to set up "ghost labs" in China while spending the majority of their time in the West where their primary lab was based. Chinese host institutions were happy to have these foreign scientists listed as part of their faculty because of the reputational remittances they provided, but local scientists (including returning Chinese scientists who had moved lock, stock and barrel back to China) felt that they were being treated like second-class citizens in their own country, even though they were the ones who were there throughout the year to mentor and advise students and lead departments.

Other scientists felt that the bias worked in the opposite direction with non-Asian expatriates feeling somewhat marginalized in

---

[26] At the same time, it is the case that my Singaporean interviewees had been the foreigners in the USA and so it is perhaps understandable that they would have been more appreciative of the American approach to diversity hiring.
[27] Bird's nest is an expensive Chinese delicacy made out of the edible nests of Asian swiftlets.

Asian universities and research organizations that they saw as privileging ethnicity and nationality over merit. One Indian scientist I interviewed in Singapore believed that Western scientists and trainees were not as invested (as non-Singaporean Asian scientists) in their lives in Singapore because of the difficulties they faced trying to secure permanent resident status in the country. Even in a country as multicultural as Singapore, Suresh felt that the Singaporean government prioritized permanent residence applications from Asian nationals over applications from people from Western countries. This led to a vicious downward cycle with Western scientists feeling less of a long-term commitment to their careers in Singapore:

> People from the West are much more likely to go back. There is no question. ... At some point, if the foreigners feel that they are second-class, they will leave. There are a lot of foreigners I talk to, mostly Westerners, who say, "Hey, the white people are an endangered species here. We are not wanted. ... They just want to promote locals, and if not locals, then Indians and Chinese, and they don't want us here and they are sidelining us." ... But it should not get to the point where they just, they all walk away.

Even as interviewees outside Singapore saw it as the most inclusive of the four Asian case countries, returnees in Singapore all seemed to think that the country's numerical diversity along ethnic lines did not equate to actual equality of treatment across different identity groups. All of these differing points of view signal how the process of balancing the priorities of inclusivity and meritocracy is not easy, especially if different groups perceive these two goals to be in conflict with one other. This is not to say that countries and research organizations – whether in Asia or elsewhere – should shy away from efforts to become more diverse. Rather, these Asian research organizations need to push for both an embrace of diversity *and* the meritocracy of ideas to be core dimensions of their scientific cultures. As one of my Taiwanese scientists lamented:

> We are too conservative. We have to be more forward. Of course, we should still preserve our identity because that is why we are different from others. But we should become more multinational if we want to build hi-tech technology. You need to get the best people from outside, not [just] from local. If you are conservative, you cannot get the best people.

Conclusion

This chapter introduced the concept of scientific remittances to highlight how returning Asian scientists bring back not just new scientific knowledge and network connections but also new norms and values about how best to engage in scientific learning and research. While these scientists are a force for cultural change in their labs and classrooms, their efforts interacted with broader societal changes occurring in their respective countries that were independently affecting young people's attitudes towards scientific knowledge and also towards science as a career choice. These societal changes could sometimes bolster scientists' change efforts but, other times, they undercut the lessons that scientists were trying to teach their students. For instance, increasing individualism in Asian societies could be helpful in creating an environment in Asian labs and classrooms where ideas can be freely debated and hierarchies are not as rigid. But young people's declining interest in science as a career is harder to combat without a large-scale governmental effort to promote Asian children's engagement with the sciences from an early age.

For larger countries like India and China, these issues are less of a problem because of the advantage that comes with their having high population numbers. Even if the proportion of young people interested in the natural sciences declines somewhat with time, there will still be enough people to drive these two countries' scientific and technological ambitions forward for the next few decades. This is the case for India, while China has a looming demographic crisis on its hands. In response, China is transforming its primary and secondary educational systems to encourage more innovative and broad-based teaching which, if successful, would allow it to reap the dividends when its young people reach the tertiary level and beyond. A shift in young people's interest away from the natural sciences is harder for smaller countries like Singapore and Taiwan, which do not have the luxury of large native populations to compensate for such shifts. Singapore is addressing this issue by relying on science-oriented international students and trainees from other countries in Asia and around the world. This will have to be the path that Taiwan adopts if it hopes to maintain its scientific reputation in the face of its declining native-born population.

Returnees in all four case countries acknowledged that their change efforts were going to take a long time to reach fruition and scale up. Some had grown frustrated with the slow pace of organizational change and had decided to stop pushing for change in the short term. Hoe Fang, who had returned to Taiwan, had decided that the smartest course of action was to resign himself to this cultural inertia until he was "senior enough" in his institute to push through more systemic changes to how research teams were formed, managed and mentored in his organization:

> When you come back, you have to withstand and you have to try to forget the good things, the good colleagues, the passion of doing science and the dedication of doing science of your American colleagues. You have to forget them. Otherwise, you will become less content. You become frustrated. [*Hoe Fang pauses*] Because you know there are much better ways of doing things. But here, you couldn't do that because you are not senior enough.

Even as Hoe Fang struck this rather depressing note, his own actions showed that he was actively trying to make a change in the scientific culture in the relatively contained arena of his lab team. He shared with me how he was trying to inculcate a norm of reading widely and of having critical discussions among his students during the weekly lab meetings he had instituted when he first returned.

Other returned scientists were more hopeful about the potential for widespread cultural change in their specific research organization in Asia. Many of them had returned to brand new research organizations that had been set up to be separate from their country's existing research and education infrastructure, precisely so that they would not be burdened by legacy cultures of patronage, acquiescence, hierarchy and/or conservatism. These new organizations allowed for the leadership to establish a fresh research culture and new research system that supported one another. Vanya, who had returned to an independent research institute in India, told me with excitement:

> There's a sense of energy that you feel that things are changing and there's a sense of awareness that you are a part of that change, even though you might be just like one little brick in a very large building. It doesn't really matter. It's a sensible[, as in palpable,] energy of change.

Vanya acknowledged that she was still on the vanguard of change in India given the elite status of her research institute, and that it would take many more years before there was widespread cultural change across all the country's research organizations. But she was hopeful that, in the long run, India's scientific sector would evolve for the better:

> You have the feeling that there is change in the air and certainly I see it in many, many things that do happen. But it's still going to be a while. It's a sense of being part of that movement. Twenty years from now, it will be a different story but, right now, it's not there yet.

In fact, many of the returnees I talked with, especially if they moved back to their birth countries, or had naturalized and become citizens in their halfway-return country, spoke of their willingness to invest their time and energy educating the next generation of Asian scientists and building the scientific culture in their research organization from the ground up. A senior scientist in Singapore, who had spent two decades in the country and was close to retirement, spoke of how much he enjoyed traveling across Asia to give lectures, run workshops and recruit young people to enroll in his Singapore university:

> I go to some Chinese university to teach and I teach them freshmen seminars [and] make sure they do oral presentations in English. It is amazing just to try to introduce the kind of teaching, that we already have, in their university. And the other thing [I do] is bring in some colleagues from the ASEAN community or going over [there] to teach their graduate students. ... I find that people really appreciate your efforts to help. That to me, as an academic, is the most important. You see the students are very enthusiastic, coming right to you; they want to come over here [to Singapore] for their attachments, internships. That is quite rewarding.

In this manner, returning scientists were beginning to impact not just the cultures in their specific research organization but also in other organizations in their country and the region at large. As the number of returning Asian scientists continues to increase, it will become even easier to observe this cultural shift in the next generation of Asian scientists being trained in Asian labs and classrooms.

9 Conclusion

Asian Scientists on the Move tells the story of how the Asian scientist migration system is undergoing change in the twenty-first century, with increasing numbers of Western-trained Asian scientists choosing to return to Asia, growing numbers of aspiring Asian scientists pursuing more of their training in Asia itself and delaying the moment of their first departure to the West, and also increasing intra-Asian mobility among both students of science and academic scientists. Through my interviews with 119 Asian-born, Western-trained scientists spread across five different countries, I show how the strategy adopted by the four Asian case countries in this book – China, India, Singapore and Taiwan – of investing heavily in upgrading their scientific research systems and higher education institutions in order to raise their relative standing within the global scientific field and to lure Western-trained Asian-born scientists back to Asia is paying dividends. More and more Western-trained Asian scientists are seeing possibilities for career success in Asia itself, while recognizing the continuing advantages of being closer to their parents and to a more familiar culture.

Upon their return, these ambitious scientists are further shifting the topography of the global scientific field. They do this in multiple intersecting ways, as I uncover through my interviews with returned Asian scientists. After they set up their new labs in Asia, their research output and research reputation boosts the standing of the Asian research organizations they join. But returnees also directly and indirectly encourage the next generation of Asian science students to pursue more of their training in Asia, where they can work for and support the research ambitions of these returned scientists. Returned scientists are actively building new research networks in their return country and across Asia to mitigate the decline in their access to Western-based networks. They are also attempting to recreate, through their scientific remittances, certain aspects of the scientific cultures they were exposed to in the West, thereby reshaping the attitudes and expanding the research ambitions of the next generation of Asian scientists. As a result of all these changes, even as Western countries, and particularly the USA, remain at the core of the global scientific field, select Asian countries are muscling their way out of the periphery of the field and ever closer to the core.

Theoretical Contributions

From a theoretical point of view, this book introduces several concepts and frameworks that can be useful for scholars and students of migration studies and science & technology studies (STS).

The first of these is the idea of an Asian scientist migration system. Rather than focusing exclusively on the popular idea of "brain circulations," which encourages a more individualistic perspective when studying the migrations of the highly skilled, this book highlights how the westward flow of Asian science students is part of a larger Asian scientist migration *system* that involves governments, universities and other research organizations in various Asian and Western countries. I show how this migration system serves as a key source of scientific human capital to research organizations in the West, because aspiring Asian scientists who move to the West for training serve as essential manpower in the labs of Western-based scientists. Asian scientists who choose to stay on and work in the West after completing their training, provide even more scientific contributions to research organizations in Western countries. In the past, the fraction of scientists who returned to Asia was enough to staff small university departments in their home country and teach the next generation of Asian science students, encouraging their best students to also head West for training. And thus, this migration cycle repeated itself for years.

But *Asian Scientists on the Move* shows how this migration system is undergoing rapid change in the early decades of the twenty-first century due to a range of factors, the most important being the heavy investment by select Asian governments to upgrade their country's scientific research systems. This increased support by select Asian governments was occurring against a backdrop of financial belt-tightening in Western countries, making the pursuit of a research career in Asia look increasingly palatable. By using a systems approach to unpack the increasing return migrations of Western-trained Asian scientists, I am also able to identify the feedback mechanisms that are instigating even further changes to migration patterns within the Asian scientist migration system.

The deconstruction of scientific cultures into their constituent elements is the second theoretical contribution of *Asian Scientists on the Move*. Scientific cultures are multifaceted sets of norms and values, shared within a given scientific community, about the socially appropriate ways to approach the practice of scientific learning, teaching and research. As opposed to imagining that there are certain key principles of science that are universally held by all scientists, a scientific cultures approach acknowledges that different scientists can hold differing views about the appropriate ways to engage with existing scientific knowledge and produce new scientific knowledge. As more and more research occurs in and through teams, the different norms that research teams internalize become increasingly important to study. Even though there has already been significant research conducted on the scientific cultures of individual labs and research organizations, this book instead uses a comparative approach to identify seven dimensions along which different scientific cultures may vary: their attitude towards scientific knowledge, their approach to scientific problem-solving, the scope of their research ambitions, the degree of autonomy granted to individual members of the scientific team (particularly students), the importance given to rank and seniority, attitudes towards difference within a team along various identity dimensions, and the team's approach to communication. I am able to make comparisons across scientific cultures because my interviewees worked in multiple labs in multiple countries over the course of their training and careers. This allowed me to compare the scientific cultures of different research labs, organizations and countries, and demonstrate how significant variation exists along each of these seven dimensions, not only between Asia and the West but also at the level of countries, universities and individual labs. I show that not all of these dimensions change in the same direction as one moves across regions, countries or organizations. Accordingly, we cannot talk about a single "Western scientific culture" or a single "Asian scientific culture."

The scientists I interviewed who returned to work in Asia are themselves further changing the scientific cultures of their new Asian workplaces. Studying and working in Western countries had a deep impact on how my interviewees envisioned themselves as practicing scientists and teachers, how they wanted to work with and supervise others in their own lab, and the relative importance they gave to questions of diversity, communication, critical thinking and egalitarianism in research settings. This affected how they taught science, and managed and mentored their research staff after they returned to Asia.

This observation is tied to the third theoretical contribution of this book – the idea of scientific remittances. This idea builds on Peggy Levitt's (1998) concept of social remittances, but focuses on the informational, reputational and cultural diffusion that occurs as a result of the brain circulations of scientists. In the past, when most Asian scientists did not return home after their training in the West, their governments still hoped to benefit from their emigration through the establishment of transnational research collaborations, transnational student mentoring relationships and overseas scientists' participation in conferences in their birth country during brief visits home. But now, as more Asian scientists are returning to work in Asia, their scientific remittances are increasing because they are now leading labs and mentoring students and staff on a daily basis. While the new knowledge and experience they bring back with them, and the improved reputational standing they provide their new Asian research organizations, are critical aspects of their scientific remittances, I demonstrate that the cultural diffusion that occurs as a result of these scientists' return is a less visible but equally important aspect of the changes they are instigating in the scientific terrain of select Asian countries. Most returnees come back as assistant professors and so they tend to limit their initial cultural change efforts to their classrooms and labs. But as they rise up the ranks within their research organizations, we should expect to see them pushing for wider changes in local research systems and scientific cultures. Even as some Asian governments fixate on the increase in citations and patents that can be linked to the return of Western-trained scientists to Asia, they should not ignore these micro-level cultural changes that have been kickstarted as a result of these return migrations, as these cultural shifts have the potential to scale up in the long run to have a discernible impact on future generations of Asian scientists.

A fourth contribution of this book relates to the conceptual framework I developed to explain the return migration decision. I organize the factors which affected my interviewees' return decision along three axes of influence – integration, obligation and ambition. In the past, the polarity of all three axes was largely aligned and heavily tilted towards the West. Scientists who felt less integrated in their Western society, and who felt obligated to care for elderly parents and grandparents back in Asia, were more inclined to return. Scientists whose professional ambitions leaned towards helping their birth country's research agenda rather than their own, were also more likely to return. But the polarity of research ambitions has shifted in recent years, with ambitious Asian scientists increasingly feeling that they can pursue their research agenda in the top research organizations in Asia itself. This shift in polarity is encouraging more Western-trained Asian scientists to return to Asia without feeling that they are making as big a professional sacrifice as earlier generations of Asian scientists in the West might have felt. And through their scientific remittances, these returned scientists may influence others (including their students) to start seeing Asia as the place to dream bigger and aim higher.

The final contribution of *Asian Scientists on the Move* relates to the work I did on how gender affects Asian women scientists' career course. The novel concept of gender shock, which I introduced in Chapter 6, captures the experience of entering a social and symbolic space where the attitudes, norms and beliefs concerning one's gender are unexpected and unfamiliar in either a positive or negative way. This conceptual innovation has the potential to contribute to gender studies more broadly and not just to the study of women in science. From my interviews, I find that the experience of negative gender shock can motivate an individual to make a gender compromise – a settling for a less-than-preferred course of action in one domain of life (whether career, marriage, etc.) – to satisfy gender-related pressures in the same domain or another one. As already observed by other STS scholars, I too find that women scientists are much more likely to make gender compromises than their male partners, though these compromises do not always involve sacrifices on the career front. Additionally, however, I observe that under certain circumstances, and especially as they gain more symbolic power in their relationships when they advance in their careers, women are able to insist that their husbands (if they are married) make more gender compromises.

Whether taken together or individually, it is my hope that the concepts and frameworks introduced in this book will be useful for scholars from a variety of fields and studying a variety of populations – not just Asian academic scientists.

Policy Recommendations

I close this book by offering seven policy suggestions derived from studying the successes and ongoing challenges experienced by my four Asian case countries, and from my analysis of the accounts of Western-trained Asian scientists in these four countries but also those who chose to stay and work in the United States. These are recommendations for the four case countries but also for other Asian (and non-Asian) countries seeking to raise their research output and relative standing in the global scientific field. I largely focus on recommendations that apply to all countries but also occasionally add on more targeted recommendations tailored to the situation faced by a specific subset of countries. I try not to rehash the changes that have already been implemented by several of my Asian case countries, such as upgrading their research infrastructures. Instead, I focus on areas which have either been given insufficient priority by states, or that might need to be reemphasized. I also restrict myself to only a general discussion of these recommendations, as offering more detailed suggestions would almost require a separate book. These recommendations are addressed to policymakers as well as research leaders who may be running research universities/institutes. Some of these recommendations require a state or research organization to make substantial financial investments, which may not be feasible for those that are small or less endowed. Other recommendations suggest changes in policy direction including funneling existing resources in new directions.

Unleash the State

My analysis of all four of my Asian case countries demonstrates the essential role that the state plays in kickstarting rapid and large-scale improvements to a country's scientific research system and its standing in the global scientific field. This requires a significant injection of state funds to upgrade both the scientific research system and higher educational sector of the country. From my study of all four case countries, it is clear that without strong state involvement no significant progress will be made on this front – or worse, a country may fall behind as other countries' governments decide to invest heavily in their national scientific research systems. A large, positive internal shock derived from a massive injection of funding is one of the few ways to instigate a change in the status quo in existing scientist migration systems and encourage more scientists to return home. We see that in the increasing prominence of China, in particular, in global research comparisons, relative to the status of North American and European countries.

At the same time, while funding increases are necessary, a unifying vision for the role of science in the future development of the country is also required from the state. Leaders at the very top of the state hierarchy need to be vocal champions of these transformational efforts, inspiring others to dream big. Public-facing national leaders must paint a picture of their country's scientific future that will inspire young people in the country to pursue the sciences, motivate native scientists to return from overseas to help build the local scientific system and convince others to join in these efforts as well. As I mentioned in Chapter 2, independent India's first prime minister, Jawaharlal Nehru, had been a champion of science and the role it could play in India's progress.[1] India's late president A. P. J. Abdul Kalam, who served as the country's eleventh President from 2002 to 2007, was a missile scientist who led the country's indigenous guided missile program before being appointed president. A much-loved figure, he travelled around the country trying to inspire young people to pursue a career in science. As Chairman of A*STAR, top civil servant Philip Yeo roamed the world in the early 2000s, recruiting top international scientists to relocate to Singapore when the country was building Biopolis and kickstarting its biomedical initiatives. Such high-level cheerleaders are essential in generating a palpable sense of excitement and contagious enthusiasm for science within a country.

Leverage all Parts of the Research Ecosystem

While vision and commitment from the state are important, government leaders need to invite in research leaders – from funding agencies, national universities, research institutes and research-oriented corporations – to collaborate in charting out a roadmap for making this vision a reality. While the state may be the key actor funding research in a country, it is critical that it does not crowd out the private sector. Otherwise, an overdependence on state funding may develop,

[1] Arnold 2000, 2013.

which was an issue identified in both Singapore and China (as mentioned in Chapter 2).[2] Countries have to guard against creating a culture of complacency among not only their scientist community but also their business community, which might assume that the state will always be there to fund research, no matter how big or small the project. It is necessary to nudge scientists to be more entrepreneurial and seek out private funding by considering the translational value of their research. Companies should also be encouraged to fund university research, in addition to their own in-house research teams, to avoid an overreliance on the state. But this requires universities to become more proactive in seeking out these private research funding partnerships and in providing the legal support structure to help their research faculty feel secure that their intellectual property (IP) will be protected. Many research universities have already set up technology transfer offices to capitalize on these research partnerships. At the national level, this also requires governments to strengthen their national IP laws, so that companies do not worry that their funded research will be stolen or copied by competitors.

University leaders should also consider how to transform their research universities into entrepreneurial hubs that foster the creation of start-ups by not just their faculty but also their more enterprising students.[3] These students and also alumni may be more naturally inclined to pursue Mode 2 (applied) research[4] than academic scientists who often choose academia as a career in order to expand our theoretical understanding of the world. Students and alumni will appreciate the infrastructural support, legal guidance and experience-based advice that their university can make available to them to scale up their business ideas. In Singapore, for instance, both NUS[5] and NTU[6] have invested heavily in establishing start-up incubation hubs on their campuses, and supporting their students to find internships at local and overseas start-ups so that they can learn firsthand how to generate business ideas, secure funding and build a company. Likewise, Tsinghua University in China has set up the "x-lab," a platform that

---

[2] Zhang et al. 2010; Kermani and Zhou 2007.
[3] Duruflé, Hellmann and Wilson 2018.
[4] Gibbons et al. 1994; Nowotny et al. 2001.
[5] Enterprise, National University of Singapore, accessed on March 25, 2021, https://enterprise.nus.edu.sg/.
[6] NTUitive, NTUitive Pte Ltd, accessed on March 25, 2021, www.ntuitive.sg/.

brings together students, faculty, alumni and investors around possible start-up ideas across fourteen different schools and departments within the university.[7] In this manner, research universities are ideally suited to encourage the entrepreneurial spirit of not just their faculty but also other members of their learning community.

Professionalize Research Administration

States and research organizations can also improve their scientific research infrastructure by streamlining and professionalizing their research-related administrative functions. These need to run smoothly and efficiently; otherwise, scientists will become frustrated with the administrative roadblocks that they continuously encounter in their research. Unfortunately, even though the administrative support role is acknowledged by many within the science hierarchy to be vital, this role is often not given sufficient respect and status within research organizations. Little thought is given to the staffing of these positions or to the career development of the individuals who fill these positions. The professionalization of research management requires administrative staff with undergraduate and graduate degrees in science to be hired so that they will be able to speak the same scientific language as practicing scientists and understand their needs and priorities better. There are always students who love science but decide that a high-paced research career is not for them. There may also be many people who do not even bother pursuing a science degree because they imagine that the only possible career for someone with a science degree is that of a scientist. To maintain a healthy pipeline of students into science, it is necessary to show the diverse range of science-related career possibilities, including in research administration, and to demonstrate how the skills and knowledge learnt from a science education can be of use in many careers and not just in academic research.

Balance Research Priorities

While applied research is useful because of its clearer revenue-generating and problem-solving potentialities, states should not shun

---

[7] Tsinghua x-lab, Tsinghua University, accessed on March 26, 2021, www.x-lab.tsinghua.edu.cn/en/.

investments in basic research. There must always be a balance between basic and translational research, and both big and small states need to develop a tolerance for research that extends our theoretical understanding of the natural world without having an immediately obvious application. Encouraging researchers to develop a portfolio of research projects – some more translational and some more basic – may help placate scientists who relish their academic freedom to pursue any research question they find intriguing, and also policymakers who want relatively quick economic returns on their investment of public funds into scientific research. Publicly setting aside a proportion of national R&D funds to be exclusively used for basic research is another way for states to signal that they still care about Mode 1 research.[8] While it is difficult to determine what the ideal proportion should be for large versus small states, or high-income versus low- or medium-income states, countries that allocate a significantly smaller part of their overall R&D budget for Mode 1 research tend to be viewed negatively by scientists. While roughly 7 percent of Taiwan's R&D budget (NTD 46 billion) was spent on basic research in 2018, South Korea allocated 14 percent of its national R&D budget to basic research, while Singapore allocated 24 percent.[9] Scientists may read a country's emphasis on basic research as a signal of its "real" commitment to science, and therefore may be less interested in working in a country where that percentage is relatively low. We saw that in the rapid departure of several superstar scientists from Singapore when the country began shifting its research emphasis towards more applied work in 2010. Governments thus need to be careful about the message they send when they push too hard in the direction of applied research, especially if this comes with the expectation of guaranteed near-term returns on their investment.

Earmarking some portion of national research funds for moonshot projects that have a much longer time horizon before they can start showing results, will avoid inadvertently encouraging scientists to only focus on incremental research questions with limited payoffs. Google's X, also called the "Moonshot Factory" is an example of a large technology organization setting aside significant resources to pursue cutting-edge research to solve critical world problems. A key

---

[8] Gibbons et al. 1994; Nowotny et al. 2001.
[9] Lin 2020.

aspect of X's operations is its celebration of both its successes and failures, because all of those endeavors still involved significant learning regardless of their outcome. Singapore's Research Innovation Enterprise 2025 plan (RIE2025) has set aside roughly 15 percent of its funds – amounting to approximately SGD 4 billion out of a total of SGD 25 billion – for "white space" research in unanticipated areas or opportunities.[10] While not quite the same as funding moonshot projects, this approach still creates funding structures that encourage a degree of spontaneity in dreaming up solutions to unexpected and complex problems.

Reform Science Education

Asian states also need to rethink how they approach the teaching of science in their schools to encourage a more inquiry- and discovery-driven engagement with scientific research and scientific methods from the earliest possible age. Rather than focusing exclusively on the memorization of scientific facts, schools need to encourage children to arrive at scientific conclusions through experimentation and trial and error as a closer reflection of the scientific process. This kind of educational reform is a massive endeavor that requires the training of a new generation of science teachers for the primary and secondary school levels, and a rewriting of the science curriculum in a country, to make them more inquiry driven. As mentioned in Chapter 8, China recently launched a nationwide reform of its school science curriculum to encourage more creative and broad-based learning. Programs should also be set up to allow interested high school science students to take university classes and spend their summers seconded to a research lab where they can see the scientific process in action in a professional setting.

At the same time as pretertiary science education needs to be reformed in many countries, tertiary science education also needs to become more integrated, so that students can uncover and appreciate the connections across different subfields. This will allow for more creative possibilities in the interstices between subfields.[11] At the tertiary level, more interdisciplinary degree programs and centers should be established, helmed by senior scientists who can bring

[10] Sharma 2021.
[11] Boden 2004; Du Sautoy 2019.

together a team of researchers drawn from a variety of specialization areas and keen to collaborate across disciplinary lines. To achieve these goals, some countries have decided to build new tertiary education institutions sitting outside the existing system of traditional universities, rather than attempting to change the system from within. This is the approach that India has taken with its establishment of the IISER campuses. While this approach allows for a more rapid rollout of new educational initiatives with a smaller and more targeted group of students, the Indian state cannot rely on these efforts if it wants to meet its goals for a nationwide economic and scientific transition.

Strengthen and Diversify the Science Pipeline

Hopefully, this book has convincingly demonstrated that, in the long run, less technologically advanced states cannot keep outsourcing all their scientific training to more advanced countries, if they want to recoup sufficient manpower to improve their research standing in the global scientific field. They may need to adopt such an outsourcing strategy during the early days of their scientific development but, in the long run, this approach will stall their scientific advancement. Hiring fully trained independent scientists from overseas also does not do enough to build a sustainable foundation of young scientific talent for the country's future. Instead, states need to consider the entire pipeline of academic science and ensure that sufficient graduate students are being trained and sufficient postdoctoral fellows recruited so that homegrown talent can be developed.

For states that cannot sustainably rely only on their own native populations to propel their scientific research ambitions, looking regionally for additional scientific talent is a viable human capital strategy. Singapore is ahead of the curve in this regard as it already draws young scientific talent from India and China, as well as from neighboring Southeast Asian countries. Other countries could follow suit by putting themselves on a path to becoming regional training hubs for aspiring scientists from neighboring countries, and also potential halfway-return destinations for scientists from the region. My interviews demonstrate that this is starting to happen with certain locations. Indian research universities, for instance, routinely draw students from neighboring South Asian countries including Nepal, Afghanistan and

Bhutan – to name the top three source countries for foreign students in India.[12]

In addition to looking to recruit foreign students and scientists, countries should also investigate if the demographics of their local scientist population match those of the national population along gender, ethnic, class, caste, religious and racial lines. If they do not, this discrepancy could be a sign that certain identity groups are being directly or indirectly cut off from pursuing science as a career. This represents an untapped (and therefore wasted) potential within the country's existing human capital base. While the interviews I conducted for this book show that growing internationalization brings in new energy and new ideas into a research organization or research lab, some countries may be overly focused on bringing in foreign students and faculty into their local scientific research system instead of investing in long-term efforts to open up opportunities for underrepresented groups from within their own native-born population. In the same way that science should no longer be viewed as a male-only pursuit, it also should not be seen as the exclusive preserve of middle- and upper-class or high-caste individuals, or of only a specific ethnicity within a country. At the tertiary level, scholarships should be offered to students from more disadvantaged or underrepresented groups who are interested in pursuing degrees in the natural sciences at local universities, and these scholarships should come with dedicated mentoring and guaranteed placements in both academic and corporate research labs so that these students can better understand the range of career possibilities available to them in the sciences.

Gender Mainstreaming Science

Continuing with the theme of strengthening the scientific pipeline, countries need to provide more support for women in science. The ongoing challenges faced by existing and aspiring Asian women scientists are a key finding of this book. It should not be the case that the only way for Asian women to rise in their chosen careers is to train and then work in Western countries. Asian governments need to be honest about the many barriers that currently exist within their scientific research systems and

[12] Department of Higher Education 2019.

scientific cultures that prevent the flourishing and advancement of local women scientists. Establishing an annual day to celebrate women in science,[13] or creating an award for high-achieving women scientists,[14] is laudable but does not actually generate any structural change that benefits *all* women pursuing scientific careers. Research leaders in a country or organization need to initiate a thorough investigation of all parts of the research environment that may be impeding women's full participation in scientific careers, including outright discrimination from male professors and peers on the basis of old-fashioned stereotypes about a woman's ability to excel in scientific research. A zero-tolerance policy against gender discrimination needs to be instituted in research organizations, and specific policies need to be put in place to support women scientists who have children or would like to start a family. At the national level, research grants won by women scientists should be automatically extended if the woman goes on maternity leave. At the organizational level, it should be standard policy that women who go on maternity leave will have their tenure clocks or contracts extended and that they will not need to undergo an annual review of their research outputs for the year when they gave birth.

The Rise of Asian Science?

Throughout this book, I have rejected the idea of an "Asian scientific culture." But there is one area where a pan-Asian scientific development is beginning to occur. This emerging phenomenon was highlighted by

---

[13] In 2015, the UN General Assembly voted to establish an international day to celebrate the role played by women and girls in science and technology. International Day of Women and Girls in Science is celebrated on February 11 annually ("International Day of Women and Girls in Science," Commemorations and Anniversaries, United Nations Educational, Scientific and Cultural Organization [UNESCO], accessed on March 23, 2021, https://en.unesco.org/commemorations/womenandgirlinscienceday).

[14] In 1998, the cosmetics company L'Oreal and UNESCO established the "International Awards L'Oréal-UNESCO for Women in Science," which are awarded to five renowned women scientists from the five main world regions each year, and come with a cash prize of EUR 100,000 for each laureate. In 2000, L'Oreal and UNESCO also established an International Rising Talents award, which recognizes fifteen younger women scientists, who each receive a EUR 15,000 prize as well as the opportunity to participate in a leadership training program (For Women in Science, accessed on March 23, 2021, www.forwomeninscience.com/).

returnees in all four countries and relates to the growing sense of competition between research organizations in Asia to recruit not only potential returnees from the West but also scientists already based in other Asian countries. While it was still rare for Western-trained Asian scientists who had returned to Asia to move laterally back to top universities in the West, there does appear to be growing mobility of academics *within* Asia, and increasing attempts by research organizations in Asian countries to recruit scientists from peer organizations in Asia.

One senior scientist in Singapore saw this as the clearest sign that the standards of Asian science were rising, as top Asian universities were not only recruiting from Western universities but also from other Asian universities:

> A concern is when my [Chinese] staff are going back to China because they really know how to recruit people back with all the big money and also the big science and also the glamor of being in the big pond versus in Singapore. I think the same thing for India too. In fact, I have one of my [Indian] colleagues ... who was offered to go back to be head of one of the research institutes in India.... Then that means that they are people of interest. In fact, some Chinese universities – they came over here and talk to our staff.... They were looking for English-speaking professors in China with experience with the West, but luckily so far, our guys preferred to stay here.

From this scientist's point of view, the high degree of interest in his research staff meant that he had done a good job hiring and mentoring these top-notch junior and mid-career scientists. As he put it, "You want to build up the department so your staff are freely movable." The increased marketability of scientists based in Asia is a sea change from just twenty years ago, when the idea of moving back to Asia tended to be read as a foolhardy move or one adopted by those lacking in ambition. Increasingly, the valence of being based in Asia – and particularly in certain parts of Asia and in top Asian research organizations – is changing. In the coming years, we should expect to see growing intra-Asian scientific mobility at the training stage as well as at all subsequent stages in Asian scientists' career course.

While the gravitational pull of the West, and particularly the USA, continues to be strong, Asia is emerging as a new gravitational center in the global scientific field. Thanks to the generous government investments and also the return migrations of Western-trained Asian scientists, newer cohorts of aspiring Asian scientists now have more training and career options available to them in Asia itself. Given the larger population base within the Asia region, this will inevitably propel the scientific progress of Asian countries in the coming years, if they are able to capitalize on and support all of this scientific talent.

Future Research

Having reached the end of this book project, I am excited to look ahead at the new research questions that working on this book has revealed. Some questions are self-evident, such as the need to continue studying how present and future generations of Asian science students engage with the question of international migration for training, and whether or not the trends I highlight in this book strengthen or not. These trends include the delay in aspiring Asian scientists' first westward departure moment to the postdoctoral training stage, as well as increasing intra-Asian mobility at the training stage and at later career stages. But there are other research questions that this book has inspired that I find equally intellectually stimulating.

First, even though I exclusively interviewed academic bioscientists for this book, my findings are likely to be applicable to other fields. Various Asian countries are investing heavily into robotics[15] and, accordingly, this might be an additional scientific field where we should expect to see significant diversification of Asian scientist training migrations as well as increasing return migrations from the West. It is unlikely, however, that parallel developments will be seen outside of STEM fields. Asian governments have tended to take a very pragmatic approach in their investments in higher education. Even though there is some shift towards emphasizing the social sciences in Asia, especially in fields (such as economics and international relations) that are viewed as useful for informing public policy, the humanities have not been as much of a priority for most Asian governments. It is unlikely that we

---

[15] "How Nations Invest in Robotics Research," International Federation of Robotics, June 5, 2020, accessed on March 22, 2021, https://ifr.org/ifr-press-releases/news/how-nations-invest-in-robotics-research.

will see a big influx of returning Asian social scientists and humanities scholars from the West, or big growth in these departments in Asian universities. But this projection of mine is ripe for testing. It would be worthwhile to disaggregate changes to the overall Asian academic migration system by discipline in order to isolate the internal and external factors that shift the gravitational pressures in some academic fields but not in others.

Another phenomenon that has yet to be adequately explored is the eastward migrations of non-Asian scientists. As I mentioned at the start of this book, I interviewed sixty-five Western life scientists who had relocated to various countries in Asia for work. Of these, seven were of Asian ethnicity and were either second- or 1.5 generation immigrants[16] in the West. The vast majority of these Western scientists were working in Singapore, but there was also a small handful of Western-born scientists I interviewed in Japan, South Korea, China (including Hong Kong) and India. I chose not to include the insights from my interviews with these Western scientists in this book so as to focus on the return migrations of Asian scientists, which vastly outnumber the eastward migrations of Western scientists. But in my next project, I hope to explore the experiences of Western scientists in Asian labs and research organizations. In Chapter 8, I touched on some of the diversity challenges faced by Asian universities and research institutes that are actively recruiting foreign faculty, and these issues were raised by many of my Western interviewees in Asia as well. Pushing to change local scientific cultures in an Asian research organization can be difficult but I found that this was even more charged when the change agent is a non-Asian foreigner. In the coming years, I hope to unpack the migration motivations and experiences of Western scientists in Asian labs to see how their decision-making varies from that of returning Asian scientists, and to explore how their migrations may signal further shifts in the topography of the global scientific field.

A third avenue for future research relates to the variations in scientific cultures that exist across Asian research organizations, and the impact that this has on individual scientists' research productivity and science students' career and migration choices. An important

---

[16] While second-generation immigrants are the locally born children of immigrants to a country, 1.5 generation immigrants arrived in the destination country as young children making their assimilation slightly easier than the experience their parents went through.

contribution of this book lies in how it differentiates between the scientific terrains in each of my case countries, rather than treating Asia as a monolith. My interviews with scientists from a range of Asian countries, and my comparative analysis of scientific cultures across multiple country sites led me to reject the idea of a single "Asian science." But there is much more work that needs to be done in this area to systematically document the multiple scientific cultures that exist across Asia. A further step would be to determine if specific cultural differences can be tied to variations in research satisfaction and productivity, as well as to show how scientific cultures evolve over time.

I am also intrigued by the increasing intra-Asian migrations I observed among my interviewees, and the ways in which they spoke of moving from the West to an Asian country other than their birth country as still a form of return. I have been asked why I count such migrations as "halfway-returns" and the simple answer is – I do so because my interviewees did. The growing competition between research organizations in mainland China, Hong Kong, Taiwan, Singapore, India and other countries in Asia to recruit from the same pool of Asian scientists also speaks to an increasing regionalization that may lead to the emergence of a shared pan-Asian identity, recognized as such by scientists of various Asian nationalities. It is an open question whether or not migration within Asia will encourage the development of a cosmopolitan Asian identity among similarly trained and professionally connected Asian-born scientists. Interviewees who trained in the USA and other Western countries often developed strong bonds of friendship with other Asian graduate students and postdoctoral fellows, regardless of their nationality, due to their shared Asian background and shared international student status. When compared with the European project of intraregional mobility and the development of a pan-European identity particularly among European youth, it is worth studying if there will be an emergence of a similar pan-Asian identity among Western-trained Asian scientists who return to Asia. While not quite the same as Merton's principle of universalism in science, it is worth asking how these returned scientists' sense of a *regional* Asian identity will interact with their particular country's nationalistic desire for a state science that supports their particular country's interests.

I end here, excited about the many research questions that still need to be answered about the rapidly shifting global scientific field, and about Asian countries and Asian scientists' place within this field. My hope is that the concepts and frameworks I developed in *Asian Scientists on the Move* will be useful to policymakers and researchers interested in the causes and consequences of these changes, to students in and of Asia who may not realize just how much has changed in the Asian scientist migration system in the last couple of decades, and finally to Asian scientists who may see themselves and their experiences in this book.

Bibliography

A*STAR. 2012. *Asia's Innovation Capital: Stepping Up Yearbook 2011/12*. Singapore: A*STAR.

Acar, Feride. 1990. "Role Priorities and Career Patterns: A Cross-Cultural Study of Turkish and Jordanian University Teachers." In *Storming the Tower: Women in the Academic World*, edited by S. Lie and V. O'Leary, 129–143. London: Kogan Page.

Ackers, Louise. 2004. "Managing Relationships in Peripatetic Careers: Scientific Mobility in the European Union." *Women's Studies International Forum* 27(3):189–201.

Ackers, Louise. 2005. "Moving People and Knowledge: Scientific Mobility in the European Union." *International Migration* 43(5):99–131.

Agarwal, Pawan. 2006. "Higher Education in India: The Need for Change." Indian Council for Research on International Economic Relations working paper, no. 180. New Delhi.

Agarwal, Pawan. 2009. *Indian Higher Education: Envisioning the Future*. New Delhi: SAGE Publications India.

Alba, Richard and Nancy Foner. 2015. *Strangers No More: Immigration and the Challenges of Integration in North America and Western Europe*. Princeton, NJ: Princeton University Press.

Alba, Richard and Victor Nee. 1997. *Remaking the American Mainstream: Assimilation and Contemporary Immigration*. Cambridge, MA: Harvard University Press.

Alexander von Humboldt Foundation. 2013. *Postdoctoral Career Paths 2.0: The Golden Triangle of Competitive Junior Investigators, Adequate Academic Systems, and Successful Careers*. Proceedings of the 7th Forum on the Internationalization of Sciences and Humanities, November 15–16, 2013, Berlin, Germany.

Ali, Syed. 2007. "'Go West Young Man': The Culture of Migration among Muslims in Hyderabad, India." *Journal of Ethnic and Migration Studies* 33(1):37–58.

Altbach, Philip G. 1998. *Comparative Higher Education: Knowledge, the University, and Development*. Hong Kong: Comparative Education Research Centre, University of Hong Kong.

Altbach, Philip G. 2005. "Higher Education in India." In *Education in India*. Volume 1, edited by S. Tewari, 244–264. New Delhi, India: Atlantic Publishers.

Altbach, Philip G. 2014. "India's Higher Education Challenges." *Asia Pacific Education Review* 15(4):503–510.

Amaral, Alberto. 2008. "Transforming Higher Education." In *From Governance to Identity*, edited by A. Amaral, I. Bleiklie and C. Musselin, 81-94. Dordrecht, the Netherlands: Springer.

Ammassari, Savina. 2009. *Migration and Development: Factoring Return into the Equation*. Newcastle upon Tyne, UK: Cambridge Scholars Publishing.

Amsden, Alice H. 1979. "Taiwan's Economic History: A Case of Etatisme and a Challenge to Dependency Theory." *Modern China* 5(3):341–379.

Appadurai, Arjun. 1990. "Disjuncture and Difference in the Global Cultural Economy." *Theory, Culture & Society* 7(2–3):295–310.

Appadurai, Arjun. 2004. "The Capacity to Aspire: Culture and the Terms of Recognition." In *Culture and Action*, edited by V. Rao and M. Walton, 59–84. Stanford, CA: Stanford Social Sciences.

Archer, Louise, Emily Dawson, Jennifer DeWitt, Amy Seakins and Billy Wong. 2015. "'Science Capital': A Conceptual, Methodological, and Empirical Argument for Extending Bourdieusian Notions of Capital beyond the Arts." *Journal of Research in Science Teaching* 52(7):922–948.

Arnold, David. 2000. *Science, Technology and Medicine in India, 1760–1947*. Cambridge, UK: Cambridge University Press.

Arnold, David. 2013. "Nehruvian Science and Postcolonial India." *Isis* 104(2):360–370.

Asia Pacific Biotech News. 2004. "Incentives for Investments in Indian Biotechnology Industry." *Asia Pacific Biotech News* 8(17):940–941. http://doi.org/10.1142/S021903030400206X.

Auriol, Laudeline. 2010. "Careers of Doctorate Holders: Employment and Mobility Patterns." Science, Technology and Industry Working Papers Series 2010/4. Paris, France: OECD Directorate for Science, Technology and Industry.

Baber, Zaheer. 1996. *The Science of Empire: Scientific Knowledge, Civilization, and Colonial Rule in India*. Albany, NY: State University of New York Press.

Bakewell, Oliver. 2014. "Relaunching Migration Systems." *Migration Studies* 2(3):300–318.

Barber, Bernard. 1962. *Science and the Social Order*. New York: Collier Books.

Basalla, George. 1967. "The Spread of Western Science." *Science* 156 (3775):611–622.

Basch, Linda, Nina Glick Schiller and Cristina S. Blanc. 1994. *Nations Unbound: Transnational Projects, Postcolonial Predicaments, and Deterritorialized Nation-States*. Amsterdam: OPA.
Basu, Paroma. 2007. "Indian Biotech's Bumpy Road." *Nature* 450:580–581.
Bauböck, Rainer. 2001. Willy Brandt Series of Working Papers in International Migration and Ethnic Relations 1/01. Malmö, Sweden: School of International Migration and Ethnic Relations, Malmö University.
Bhagwati, Jagdish and Koichi Hamada. 1974. "The Brain Drain, International Integration of Markets for Professionals and Unemployment: A Theoretical Analysis. *Journal of Development Economics* 1(1):19–42.
Bhatt, Amy. 2018. *High-Tech Housewives: Indian IT Workers, Gendered Labor, and Transmigration*. Seattle, WA: University of Washington Press.
Bhattacharya, Ananya. 2018. "India Wants NRI Scientists to Come Home – But Where's the R&D Money?" *Quartz India*, January 29. Accessed March 10, 2021. https://qz.com/india/1191281/economic-survey-2018-india-wants-nri-scientists-to-come-home-but-wheres-the-rd-money/.
Biotechnology and Biological Science Research Council. 2013. *The Age of Bioscience: Strategic Plan*. London, UK: Research Councils UK.
Bobb, Dilip. 1977. "US Govt to Bar Entry of All Trained Medical Personnel into US from Every Country." *India Today*, February 15, updated March 20, 2015. Accessed January 23, 2021. www.indiatoday.in/magazine/indiascope/story/19770215-us-govt-to-bar-entry-of-all-trained-medical-personnel-into-us-from-every-country-818772-2015-03-09.
Boden, Margaret A. 2004 (1990). *The Creative Mind: Myths and Mechanisms*. London, UK: Routledge.
Boden, Margaret A. 2016. *AI: Its Nature and Future*. Oxford: Oxford University Press.
Bourdieu, Pierre. 1977. *Outline of a Theory of Practice*. Cambridge, UK: Cambridge University Press.
Bourdieu, Pierre. 1986. "The Forms of Capital." In *Handbook of Theory of Research for the Sociology of Education*, edited by J. Richardson, 241–258. Westport, CT: Greenwood Press.
Bourdieu, Pierre. 1988. *Homo Academicus*. Stanford, CA: Stanford University Press.
Bourdieu, Pierre. 1993. *Sociology in Question*. London: Sage Publications.
Bourdieu, Pierre. 2005. *The Social Structures of the Economy*. Cambridge, UK: Polity Press.
Bourdieu, Pierre and Loïc J. D. Wacquant. 1992. *An Invitation to Reflexive Sociology*. Chicago, IL: University of Chicago Press.
Bovenkerk, Frank. 1974. *The Sociology of Return Migration: A Bibliographic Essay*. Dordrecht, the Netherlands: Springer.

Bueskens, Petra. 2018. *Modern Motherhood and Women's Dual Identities: Rewriting the Sexual Contract*. New York: Routledge.

Business Today India. 2019. "Growth in R&D Expenditure to Be Targeted 2% of GDP by 2022: PM Economic Panel." Business Today India. July 24. Accessed on June 26, 2021. www.businesstoday.in/latest/economy-politics/story/growth-in-rd-expenditure-to-be-targeted-2-of-gdp-by-2022-pm-economic-panel-217740-2019-07-24.

Campion, Patricia and Wesley Shrum. 2004. "Gender and Science in Development: Women Scientists in Ghana, Kenya, and India." *Science, Technology, & Human Values* 29(4):459–485.

Cantwell, Brendan. 2011. "Academic In-sourcing: International Postdoctoral Employment and New Modes of Academic Production." *Journal of Higher Education Policy and Management* 33(2):101–114.

Carlson, Sören. 2013. "Becoming a Mobile Student: A Processual Perspective on German Degree Student Mobility." *Population, Space and Place* 19(2):168–180.

Cassarino, Jean-Pierre. 2004. "Theorising Return Migration: The Conceptual Approach to Return Migrants Revisited." *International Journal on Multicultural Societies* 6(2):253–279.

Cerna, Lucie and Meng-Hsuan Chou. 2014. "The Regional Dimension in the Global Competition for Talent: Lessons from Framing the European Scientific Visa and Blue Card." *Journal of European Public Policy* 21(1):76–95.

Cerna, Lucie and Mathias Czaika. 2021. "Rising Stars in the Global Race for Skill? A Comparative Analysis of Brazil, India, and Malaysia." *Migration Studies* 9(1):21–46.

Chakma, Justin, Hassan Masum, Kumar Perampaladas et al. 2010. "Case Study: India's Billion Dollar Biotech." *Nature Biotechnology* 28(8):783.

Chakraborty, Chiranjib and Govindasamy Agoramoorthy. 2010. "A Special Report on India's Biotech Scenario: Advancement in Biopharmaceutical and Health Care Sectors." *Biotechnology Advances* 28(1):1–6.

Chakravarthy, Radha. 1986. "Productivity of Indian Women Scientists." *Productivity* 27(3):259–269.

Chan, Yiu Lin. 2006. "The Buzz in Singapore's Biotech Industry." *Asia-Pacific Biotech News* 10(8):407–412.

Chanda, Rupa. 2015. "Internationalisation of Higher Education: Students and Institutional Mobility." In *India Higher Education Report 2015*, edited by N. V. Varghese and G. Malik, 431–457. New Delhi: Routledge India.

Chang, Ai-Lien. 2003. "Biomedical Research Hub Launched." *Straits Times*, October 30, H6.

Chang, Chen-Wei and Wang-Ching Shaw. 2016. "Expanding Higher Education in Taiwan: The Case of Doctoral Education." *Higher Education Studies* 6(1):1–14.

Chang, Kyung-Sup. 2014. "Individualization without Individualism: Compressed Modernity and Obfuscated Family Crisis in East Asia." In *Transformation of the Intimate and the Public in Asian Modernity*, edited by E. Ochiai and L. A. Hosoya, 37–62. Leiden, the Netherlands: Brill.

Chang, Shirley L. 1992. "Causes of Brain Drain and Solutions: The Taiwan Experience." *Studies in Comparative International Development* 27(1):27–43.

Chang, Zengyi. 2009. "The CUSBEA Program: Twenty Years After." *IUBMB Life* 61(6):555–565.

Chaturvedi, Sachin. 2005. "Dynamics of Biotechnology Research and Industry in India: Statistics, Perspectives and Key Policy Issues." Statistical Analysis of Science, Technology and Industry STI Working Paper Series 2005/6. Paris: OECD.

Chaurasia, Anurag. 2016. "Stop Teaching Indians to Copy and Paste." *Nature* 534:591.

Cheam, Jessica. 2005. "Bio-Mass." Straits Times (Singapore), August 13.

Chen, Chei-Hsiang. 2005. "Biotechnology Industry in Taiwan: Asian Leader, Global Partner!" *Asia-Pacific Biotech News* 9(18):959–967.

Chen, Chung-Jen and Chin-Chen Huang. 2004. "A Multiple Criteria Evaluation of High-Tech Industries for the Science-Based Industrial Park in Taiwan." *Information & Management* 41(7):839–851.

Chen, Qiongqiong. 2015. "Globalization and Transnational Academic Mobility: The Experiences of Chinese Academic Returnees." PhD dissertation, State University of New York.

Chen, Yu-Jie and Jerome A. Cohen. 2020. "Why Does the WHO Exclude Taiwan?" Council on Foreign Relations, April 9. Accessed on July 12, 2021. www.cfr.org/in-brief/why-does-who-exclude-taiwan.

Cheng, Ling-Fang. 2010. "Why Aren't Women Sticking with Science in Taiwan." *Kaohsiung Journal of Medical Sciences* 26(6):S28–34.

Chittoor, Raveendra, Sougata Ray, Preet S. Aulakh et al. 2008. "Strategic Responses to Institutional Changes: 'Indigenous Growth' Model of the Indian Pharmaceutical Industry." *Journal of International Management* 14(3):252–269.

Choi, Susanne and Yinni Peng. 2016. *Masculine Compromise: Migration, Family, and Gender in China*. Berkeley, CA: University of California Press.

Chompalov, Ivan. 2006. "Birds of Passage: Patterns of Brain Drain from Bulgaria before and after the Transition to Democracy." *Sociological Viewpoints* 22(2):41–56.

Choo, Chun Wei, Pierette Bergeron, Brian Detlor et al. 2008. "Information Culture and Information Use: An Exploratory Study of Three Organizations." *Journal of the American Society for Information Science and Technology* 59(5):792–804.

Chou, Prudence C. 2014. "Education in Taiwan: Taiwan's Colleges and Universities." Taiwan-US Quarterly Analysis Series. Brookings. November 12. Accessed on July 9, 2021. www.brookings.edu/opinions/education-in-taiwan-taiwans-colleges-and-universities.

Choudaha, Rahul and Li Chang. 2012. *Trends in International Student Mobility*. World Education Services Research Report 1. New York.

Christensen, Tom. 2011. "University Governance Reforms: Potential Problems of More Autonomy?" *Higher Education* 62(4):503–517.

Cohen, Nir. 2009. "Come Home and Be Professional: Ethno-nationalism and Economic Rationalism in Israel's Return Migration Strategy." *Immigrants & Minorities* 27(1):1–28.

Commission of the European Communities (CEC). 2001. *A Mobility Strategy for the European Research Area*. Brussels, Belgium: CEC.

Constant, Amelie and Douglas S. Massey. 2002. "Return Migration by German Guestworkers: Neoclassical versus New Economic Theories." *International Migration* 40(4):5–38.

Copeland, Neal and Nancy Jenkins. 2009. "Science in Singapore – Aiming High for Biomedical Research." *CCR Connections* 3(2).

Creese, Gillian, Isabel Dyck and Arlene T. McLaren. 2011. "The Problem Of 'Human Capital': Gender, Place and Immigrant Household Strategies of Reskilling in Vancouver." In *Gender, Generations and the Family in International Migration*, edited by A. Kraler, E. Kofman, M. Kohli and C. Schmoll, 141–162. Amsterdam: Amsterdam University Press.

Cummings, Williams K. 2011. "The Rise of Asian Research Universities: Focus on the Context." In *The Changing Academic Profession in Asia: Contexts, Realities and Trends*, edited by M. J. Finklestein, 57–78. Hiroshima, Japan: Research Institute for Higher Education.

Cyranoski, David. 2001. "Building a Biopolis." *Nature* 412:370–371.

Cyranoski, David. 2016. "Science Wins in Five-Year Plan." *Nature* 531:424–425.

Davis, Phil. 2011. "Paying for Impact: Does the Chinese Model Make Sense?" The Scholarly Kitchen, Society for Scholarly Publishing, April 7. Accessed on February 15, 2021. https://scholarlykitchen.sspnet.org/2011/04/07/paying-for-impact-does-the-chinese-model-make-sense/.

de Beauvoir, Simone. 1989 (1949). *The Second Sex*. New York: Vintage.

De Haas, Hein. 2010. "Migration and Development: A Theoretical Perspective." *International Migration Review* 44(1):227–264.

Dear, Peter. 2001. *Revolutionizing the Sciences: European Knowledge and Its Ambitions, 1500–1700*. Princeton, NJ: Princeton University Press.

Department of Biotechnology. 2012. *Biotechnology Landscape in India*. New Delhi, India: DBT, Government of India.

Department of Higher Education. 2019. *All India Survey on Higher Education 2018–2019*. New Delhi, India: Ministry of Human Resource Development.

Dewey, John. 2004 (1916). *Democracy and Education: An Introduction to the Philosophy of Education*. Mineola, NY: Dover Publications.

Dhawan, Jyotsna, Rajesh S. Gokhale and Inder M. Verma. 2005. "Bioscience in India: Times Are Changing." *Cell* 123(5):743–745.

Dhingra, Pawan. 2007. *Managing Multicultural Lives: Asian American Professionals and the Challenge of Multiple Identities*. Stanford, CA: Stanford University Press.

Dikötter, Frank. 2016. *The Cultural Revolution: A People's History, 1962–1976*. London: Bloomsbury.

Dillon, Niall. 2003. "The Postdoctoral System under the Spotlight." *EMBO Reports* 4(1):2–4.

Disease Models & Mechanisms. 2012. "Synergy in Science: An Interview with Neal Copeland and Nancy Jenkins." *Disease Models and Mechanisms* 5(6):713–717.

Docquier, Frédéric, B. Lindsay Lowell and Abdeslam Marfouk. 2008. "A Gendered Assessment of the Brain Drain." Policy Research Working Paper, no. 4613, World Bank, Washington, DC.

Docquier, Frédéric and Hillel Rapoport. 2012. "Globalization, Brain Drain and Development." *Journal of Economic Literature* 50(3):681–730.

Dodgson, Mark, John Mathews, Tim Kastelle et al. 2008. "The Evolving Nature of Taiwan's National Innovation System: The Case of Biotechnology Innovation Networks." *Research Policy* 37:430–445.

Donato, Katherine M. and Catalina Amuedo-Dorantes. 2020. "The Legal Landscape of U.S. Immigration: An Introduction." *RSF: The Russell Sage Foundation Journal of the Social Sciences* 6(3):1–16.

Du Sautoy, Marcus. 2019. *The Creativity Code: Art and Innovation in the Age of AI*. Cambridge, MA: Harvard University Press.

Duruflé, Gilles, Thomas Hellmann and Karen Wilson. 2018. "Catalysing Entrepreneurship in and around Universities." *Oxford Review of Economic Policy* 34(4):615–636.

Ecklund, Elaine and Anne E. Lincoln. 2016. *Failing Families, Failing Science: Work-Family Conflict in Academic Science*. New York: NYU Press.

Eckstein, Susan and Adil Najam, eds. 2013. *How Immigrants Impact Their Homelands*. Durham, NC: Duke University Press.

Economist. 2013. "Looks Good on Paper." September 28. Accessed on February 16, 2021. www.economist.com/china/2013/10/03/looks-good-on-paper.

Economist. 2016. "Schrödinger's Panda." June 4. Accessed on February 16, 2021. www.economist.com/science-and-technology/2016/06/02/schrodingers-panda.

Elman, Benjamin A. 2006. *A Cultural History of Modern Science in China*. Cambridge, MA: Harvard University Press.

Epstein, Jeremy. 2016. "Reflections of an NSF Program Officer." *IEEE Security & Privacy* 14(2). doi: 10.1109/MSP.2016.45.

Ezrahi, Yaron. 1990. *The Descent of Icarus: Science and the Transformation of Contemporary Democracy*. Cambridge, MA: Harvard University Press.

Faist, Thomas, Margit Fauser and Eveline Reisenauer. 2013. *Transnational Migration*. Cambridge, UK: Polity Press.

Ferlie, Ewan, Christine Musselin and Gianluca Andresani. 2008. "The Steering of Higher Education Systems: A Public Management Perspective." *Higher Education* 56(3):325–348.

Findlay, Allan M. 2010. "An Assessment of Supply and Demand-Side Theorizations of International Student Mobility." *International Migration* 49(2):162–190.

Finn, Michael G. and Leigh Ann Pennington. 2018. *Stay Rates of Foreign Doctorate Recipients from U.S. Universities, 2013*. Oak Ridge, TN: Oak Ridge Institute for Science and Education.

Fischer, Michael M. J. 2013. "Biopolis: Asian Science in the Global Circuitry." *Science, Technology & Society* 18(3):381–406.

Fisher, Doug. 2011. "First Person: Life as a NSF Program Director." *The CCC Blog*, August 24. Accessed on March 7, 2021. https://cccblog.org/2011/08/24/first-person-life-as-a-nsf-program-director/.

Fitzgerald, David S. 2013. "Immigrant Impacts in Mexico: A Tale of Dissimilation." In *How Immigrants Impact their Homelands*, edited by S. E. Eckstein and A. Najam, 114–137. Durham, NC: Duke University Press.

Fox, Mary F. 1999. "Women and Scientific Careers." In *Handbook of Science, Technology, and Society*, edited by Sheila Jasanoff, Gerald Markle, James Peterson et al., 205–223. Newbury Park, CA: Sage.

Franklin, Sarah. 1995. "Science as Culture, Cultures of Science." *Annual Review of Anthropology* 24:163–184.

Franzoni, Chiara, Giuseppe Scellato and Paula Stephan. 2012. "Foreign-Born Scientists: Mobility Patterns for 16 Countries." *Nature Biotechnology* 30(12):1250–1253.

Frew, Sarah E., Rahim Rezaie, Stephen M. Sammut et al. 2007. "India's Health Biotech Sector at a Crossroads." *Nature Biotechnology* 25(4):403–417.
Frew, Sarah E., Stephen M. Sammut, Alysha F. Shore et al. 2008. "Chinese Health Biotech and the Billion-Patient Market." *Nature Biotechnology* 26(1):37–53.
Fu, Dai-Wie and Hsiu-Yun Wang. 1996. "Women Scientists in Taiwan and Their Current Situation." *Taiwan: A Radical Quarterly in Social Studies* 22:1–58. (In Chinese.)
Furnham, Adrian and Stephen Bochner. 1986. *Culture Shock: Psychological Reactions to Unfamiliar Environments*. London, UK: Methuen & Co. Ltd.
Gallagher, Michael. 2013. "Postdocs and Changing Researcher Career Paths." In *Postdoctoral Career Paths 2.0: The Golden Triangle of Competitive Junior Investigators, Adequate Academic Systems, and Successful Careers*, 6–16. Berlin, Germany: Alexander von Humboldt Foundation.
Ganguli, Ian. 2014. "Scientific Brain Drain and Human Capital Formation after the End of the Soviet Union." *International Migration* 52(5):95–110.
Garg, Kailash C. and Suresh Kumar. 2014. "Scientometric Profile of Indian Scientific Output in Life Sciences with a Focus on the Contributions of Women Scientists." *Scientometrics* 98(3):1771–1783.
Gaw, Kevin F. 2000. "Reverse Culture Shock in Students Returning from Overseas." *International Journal of Intercultural Relations* 24(1):83–104.
Ghosh, Deepa. 2011. "Eve Teasing: The Role of the Patriarchal System of the Society." *Journal of the Indian Academy of Applied Psychology* 37:100–107.
Ghosh, Jaideep and Avinash Kshitij. 2016. "Higher Education in Basic Science and Socioeconomic Characteristics of Students' Life in India: An Exploratory Study." *Social Indicators Research* 125(1):311–337.
Gibbons, Michael, Camille Limoges, Helga Nowotny et al. 1994. *The New Production of Knowledge: The Dynamics of Science and Research in Contemporary Societies*. London: Sage.
Gibson, John and David McKenzie. 2011. "The Microeconomic Determinants of Emigration and Return Migration of the Best and Brightest: Evidence from the Pacific." *Journal of Development Economics* 95(1):18–29.
Gieryn, Thomas F. 1983. "Boundary-Work and the Demarcation of Science from Non-science: Strains and Interests in Professional Ideologies of Scientists." *American Sociological Review* 48(6):781–795.
Glick Schiller, Nina, Linda Basch and Cristina S. Blanc. 1992. "Transnationalism: A New Analytic Framework for Understanding Migration." *Annals of the New York Academy of Sciences* 645:1–24.

Glick Schiller, Nina, Linda Basch and Cristina S. Blanc. 1995. "From Immigrant to Transmigrant: Theorizing Transnational Migration." *Anthropological Quarterly* 68(1):48–63.
Goodall, Simon, Bart Janssens, Kim Wagner et al. 2006. "The Promise of the East: India and China as R&D Options." *Bioentrepreneur* (July 25). doi: 10.1038/bioent910.
Grubel, Herbert G. and A. D. Scott. 1966. "The Immigration of Scientists and Engineers to the United States, 1949–61." *Journal of Political Economy* 74(4):368–378.
Grueber, Martin and Tim Studt. 2013. *2014 Global R&D Funding Forecast. R&D Magazine*, December. Accessed on March 10, 2021. www.battelle.org/docs/default-source/misc/2014-rd-funding-forecast.pdf?sfvrsn=2.
Guarnizo, Luis E. 1997. "The Emergence of a Transnational Social Formation and the Mirage of Return Migration among Dominican Transmigrants." *Identities* 4(2):281–322.
Gupta, Namrata. 2007a. "Indian Women in Doctoral Education in Science and Engineering." *Science, Technology, & Human Values* 32(5):507–533.
Gupta, Namrata. 2007b. "Women Research Scholars in IITs: Impact of Social Milieu and Organisational Environment." *Sociological Bulletin* 56(1):23–45.
Gupta, Namrata. 2016. "Perceptions of the Work Environment: The Issue of Gender in Indian Scientific Research Institutes." *Indian Journal of Gender Studies* 23(3):437–466.
Gupta, Namrata. 2020. *Women in Science and Technology: Confronting Inequalities*. New Delhi, India: Sage Publications.
Gupta, Namrata and A. K. Sharma. 2002. "Women Academic Scientists in India." *Social Studies of Science* 32(5/6):901–915.
Gupta, Namrata and A. K. Sharma. 2003. "Patrifocal Concerns in the Lives of Women in Academic Science: Continuity of Tradition and Emerging Challenges." *Indian Journal of Gender Studies* 10(2):279–305.
Hack, Karl and Jean-Louis Margolin, eds. 2010. *Singapore from Temasek to the 21st Century: Reinventing the Global City*. Singapore: NUS Press.
Hamamura, Takeshi. 2011. "Are Cultures Becoming Individualistic? A Cross-Temporal Comparison of Individualism–Collectivism in the United States and Japan." *Personal and Social Psychology Review* 16(1):3–24.
Hambley, Connie J. 2010. "Catching the Wave in China." *Nature Biotechnology* 28(12):1309–1310.
Hamilton, Robert V., Connie L. McNeely and Wayne D. Perry. 2012. "Natural Sciences Doctoral Attainment by Foreign Students at US

Universities." George Mason University School of Public Policy Research Paper, no. 2012-06, George Mason University, Washington, DC.

Harvey, William S. 2009. "British and Indian Scientists in Boston Considering Returning to Their Home Countries." *Population, Space and Place* 15(6):493–508.

Hegarty, Niall. 2014. "Where We Are Now: The Presence and Importance of International Students to Universities in the United States." *Journal of International Students* 4(3):223–235.

Heney, Paul. 2020. "Global R&D Funding Forecast: Special Mid-year Update, Part 1." *R&D World*, August 19. Accessed on March 26, 2021. www.rdworldonline.com/global-rd-funding-forecast-special-mid-year-update-part-1/.

Hepeng, Jia. 2011. "China Eyes Western Biologics." *Nature Biotechnology* 29(11):960.

Hercog, Metka and Md Zakaria Siddiqui. 2014. "Experiences in the Host Countries and Return Plans: The Case Study of Highly Skilled Indians in Europe." In *Indian Skilled Migration and Development*, edited by G. Tejada, U. Bhattacharya, B. Khadria et al., 213–235. New Delhi, India: Springer India.

Hochschild, Arlie R. 1989. *The Second Shift: Working Parents and the Revolution at Home*. New York: Viking.

Homma, Miwako K., Reiko Motohashi and Hisako Ohtsubo. 2013. "Maximizing the Potential of Scientists in Japan: Promoting Equal Participation for Women Scientists through Leadership Development." *Genes to Cells* 18(7):529–532.

Hondagneu-Sotelo, Pierrette. 1994. *Gendered Transitions: Mexican Experiences of Immigration*. Berkeley, CA: University of California Press.

Hsu, Yeou-Geng, Joseph Z. Shyu and Gwo-Hshing Tzeng. 2005. "Policy Tools on the Formation of New Biotechnology Firms in Taiwan." *Technovation* 25(3):281–292.

Hung, Fan-sing. 2010. "Intention of Students in Less Developed Cities in China to Opt for Undergraduate Education Abroad: Does This Vary as Their Perceptions of the Attractions of Overseas Study Change?" *International Journal of Educational Development* 30(2):213–223.

Hunter, Rosalind S., Andrew J. Oswald and Bruce G. Charlton. 2009. "The Elite Brain Drain." *The Economic Journal* 119(June):F231–F251.

Hyatt, Susan B., Boone W. Shear and Susan Wright, eds. 2015. *Learning under Neoliberalism: Ethnographies of Governance in Higher Education*. New York: Berghahn Books.

Ioannidis, John P. A. 2004. "Global Estimates of High-Level Brain Drain and Deficit." *Federation of American Societies for Experimental Biology Journal* 18(9):936–939.

Ip, Nancy Y. 2011. "Career Development for Women Scientists in Asia." *Neuron* 70(6):1029–1032.

Iredale, Robyn and Fei Guo. 2003. "The Transforming Role of Skilled and Business Returnees: Taiwan, China and Bangladesh." Unpublished manuscript. University of Wollongong, Centre for Asia Pacific Social Transformation Studies (CAPSTRANS).

Jacob, Margaret C. 1997. *Scientific Culture and the Making of the Industrial West*. Oxford: Oxford University Press.

Jasanoff, Sheila. 2005. *Designs on Nature: Science and Democracy in Europe and the United States*. Princeton, NJ: Princeton University Press.

Jayaraman, K. S. and Subhra Priyadarshini. 2012. "Poor Teaching Bane of Indian Science." *Nature India*. January 13. doi: 10.1038/nindia.2012.1.

Kandel, William and Douglas S. Massey. 2002. "The Culture of Mexican Migration: A Theoretical and Empirical Analysis." *Social Forces* 80(3):981–1004.

Kapur, Devesh and John McHale. 2005. *Give Us Your Best and Brightest*. Washington, DC: Center for Global Development.

Katz, Eliakim and Oded Stark. 1986. "Labor Migration and Risk Aversion in Less Developed Countries." *Journal of Labor Economics* 4(1):134–149.

Keller, Evelyn F. 2009. "The Anomaly of a Woman in Physics." In *Women, Science, and Technology*, edited by M. Wyer, M. Barbercheck, D. Giesman et al., 23–30. New York: Routledge.

Kermani, Faiz and Yibing Zhou. 2007. "China Commits Itself to Biotech in Healthcare." *Drug Discovery Today* 12(13/14):501–503.

Khadria, Binod. 2015. "Higher Education and International Migration." In *India Higher Education Report 2015*, edited by N. V. Varghese and G. Malik, 275–304. New Delhi, India: Routledge India.

King, David A. 2004. "The Scientific Impact of Nations: What Different Countries Get for Their Research Spending." *Nature* 430(6997):311–316.

King, Russell and Parvati Raghuram. 2013. "International Student Migration: Mapping the Field and New Research Agendas." *Population, Space and Place* 19(2):127–137.

King, Russell and Gunjan Sondhi. 2018. "International Student Migration: A Comparison of UK and Indian Students' Motivations for Studying Abroad." *Globalisation, Societies and Education* 16(2):176–191.

Knight, Jane. 2008. *Higher Education in Turmoil: The Changing World of Internationalization*. Rotterdam, the Netherlands: Sense Publishers.

Koh, Winston T. H. and Poh Kam Wong. 2005. "Competing at the Frontier: The Changing Role of Technology Policy in Singapore's Economic Strategy." *Technological Forecasting and Social Change* 72(3):255–285.

Kotter, John P. 1996. *Leading Change*. Boston, MA: Harvard Business Review Press.

Kõu, Anu and Ajay Bailey. 2014. "'Movement Is a Constant Feature in My Life': Contextualising Migration Processes of Highly Skilled Indians." *Geoforum* 52:113–122.

Krautwurst, Udo. 2014. *Culturing Bioscience: A Case Study in the Anthropology of Science*. Toronto, Canada: University of Toronto Press.

Krishna, V. V. and Sohan P. Sha. 2015. "Building Science Community by Attracting Global Talents: The Case of Singapore Biopolis." *Science, Technology, & Society* 20(3):389–413.

Krishna Raj, Maithreyi. 1991. *Women and Science: Selected Essays*. Bombay, India: Himalaya Publishing House.

Kuhn, Thomas. 1962. *Structure of Scientific Revolutions*. Chicago, IL: University of Chicago Press.

Kulikoff, Allan. 1986. "Migration and Cultural Diffusion in Early America, 1600–1860." *Historical Methods* 19(4):153–169.

Kumar, Neelam. 2001. "Gender and Stratification in Science: An Empirical Study in the Indian Setting." *Indian Journal of Gender Studies* 8(1):51–67.

Kuznets, Paul W. 1988. "An East Asian Model of Economic Development: Japan, Taiwan and South Korea." *Economic Development and Cultural Change* 36(3):S11–S43.

Kwa, Chong Guan, Derek Heng, Peter Borschberg and Tan Tai Yong. 2019. *Seven Hundred Years: A History of Singapore*. Singapore: Marshall Cavendish.

Lai, Dawn. 2007. *Growth of Research and Development in Singapore: 2000–2005*. Statistics Singapore Newsletter, March. Singapore: Department of Statistics.

Lai, Si Tsui-Auch. 2004. "Bureaucratic Rationality and Nodal Agency in a Developmental State: The Case of State-Led Biotechnology Development in Singapore." *International Sociology* 19(4):451–477.

Lan, Flora, Katherine Hale and Emilda Rivers. 2015. "Immigrants' Growing Presence in the US Science and Engineering Workforce: Education and Employment Characteristics in 2013." *InfoBrief*, September. National Center for Science and Engineering Statistics, NSF 15-328.

Lancet. 2015. "China's Medical Research Integrity Questioned." 385 (9976):1365.

Langer, Eric and Eliza Y. Zhou. 2007. "Biotechnology and Life Sciences Education in China." *BioPharm International* 20(11):2–5.

Laoire, Caitríona N. 2008. "'Settling Back'? A Biographical and Life-Course Perspective on Ireland's Recent Return Migration." *Irish Geography* 41(2):195–210.

Latour, Bruno. 1987. *Science in Action: How to Follow Scientists and Engineers through Society*. Cambridge, MA: Harvard University Press.

Latour, Bruno and Steve Woolgar. 1979. *Laboratory Life: The Construction of Scientific Facts*. Princeton, NJ: Princeton University Press.

Le, Thanh. 2008. "'Brain Drain' or 'Brain Circulation': Evidence from OECD's International Migration and R&D Spillovers." *Scottish Journal of Political Economy* 55(5):618–636.

Lee, Bryan. 2005. "A*Star's Million-$ Stars." *Straits Times*, February 1.

Lee, D.-T. 2013. "Reviving Globally Stagnant Postdoctoral Careers." In *Postdoctoral Career Paths 2.0: The Golden Triangle of Competitive Junior Investigators, Adequate Academic Systems, and Successful Careers*, 17–19. Berlin: Alexander von Humboldt Foundation.

Lee, Kong-Ju-Bock. 2010. "Women in Science, Engineering and Technology (SET) in Korea: Improving Retention and Building Capacity." *International Journal of Gender, Science and Technology* 2(2):235–248.

Lee, Michael H. and Saravanan Gopinathan. 2008. "University Restructuring in Singapore: Amazing or a Maze?" *Policy Futures in Education* 6(5):569–588.

Lenoir, Remi. 2006. "Scientific Habitus: Pierre Bourdieu and the Collective Individual." *Theory, Culture & Society* 23(6):25–43.

Levin, Richard C. 2010. "Top of the Class: The Rise of Asia's Universities." *Foreign Affairs* 89(3):63–75.

Levitt, Peggy. 1998. "Social Remittances: Migration Driven Local-Level Forms of Cultural Diffusion." *International Migration Review* 32(4):926–948.

Levitt, Peggy and Deepak Lamba-Nieves. 2011. "Social Remittances Revisited." *Journal of Ethnic and Migration Studies* 37(1):1–22.

Lewin, Arie Y. and Xing Zhong. 2013. "The Evolving Diaspora of Talent: A Perspective on Trends and Implications for Sourcing Science and Engineering Work." *Journal of International Management* 19(1):6–13.

Ley, David and Audrey Kobayashi. 2005. "Back to Hong Kong: Return Migration or Transnational Sojourn?" *Global Networks* 5(2):111–127.

Leydesdorff, Loet and Ping Zhou. 2005. "Are the Contributions of China and Korea Upsetting the World System of Science?" *Scientometrics* 63(3):617–630.

Li, F. L. N., Allan M. Findlay, A. J. Jowett et al. 1996. "Migrating to Learn and Learning to Migrate: A Study of the Experiences and Intentions of International Student Migrants." *International Journal of Population Geography* 2(1):51–67.

Li, Mei and Mark Bray. 2007. "Cross-Border Flows of Students for Higher Education: Push-Pull Factors and Motivations of Mainland Chinese Students in Hong Kong and Macau." *Higher Education* 53(6):791–818.

Li, Wei, Shengnan Zhao, Zheng Lu et al. 2018. "Student Migration: Evidence from Chinese Students in the US and China." *International Migration* 57(3):334–353.

Libaers, Dirk P. 2007. "Role and Contribution of Foreign-Born Scientists and Engineers to the Public U.S. Nanoscience and Technology Research Enterprise." *IEEE Transactions on Engineering Management* 54(3):423–432.

Lin, Chia-nan. 2020. "Nation Spent 3.4% of GDP on Tech R&D." *Taipei Times*, November 9. Accessed on March 26, 2021. www.taipeitimes.com/News/taiwan/archives/2020/11/09/2003746622.

Liu, Chung-Yuan. 1993. "Government's Role in Developing a High-Tech Industry: The Case of Taiwan's Semiconductor Industry." *Technovation* 13(5):299–309.

Loder, Natasha. 1999. "Gender Discrimination 'Undermines Science.'" *Nature* 402:337.

Luthra, Renee, Thomas Soehl and Roger Waldinger. 2018. "Reconceptualizing Context: A Multilevel Model of the Context of Reception and Second-Generation Educational Attainment." *International Migration Review* 52(3):898–928.

Mabogunje, Akin L. 1970. "Systems Approach to a Theory of Rural-Urban Migration." *Geographical Analysis* 2(1):1–18.

Manning, Kimberley E. and Felix Wemheuer, eds. 2011. *Eating Bitterness: New Perspectives on China's Great Leap Forward and Famine*. Vancouver, Canada: University of British Columbia Press.

Manthorpe, Jonathan. 2005. *Forbidden Nation: The History of Taiwan*. New York: Palgrave Macmillan.

Marginson, Simon. 2011. "Higher Education in East Asia and Singapore: Rise of the Confucian Model." *Higher Education* 61:587–611.

Marginson, Simon. 2021. "What Drives Global Science? The Four Competing Narratives." *Studies in Higher Education*. doi: 10.1080/03075079.2021.1942822.

Marginson, Simon, Russell Tytler, Brigid Freeman and Kelly Roberts. 2013. *STEM: Country Comparisons; International Comparisons of Science, Technology, Engineering and Mathematics (STEM) Education*. Report for the Australian Council of Learned Academies. Melbourne, Australia: Australian Council of Learned Academies.

Mathews, John A. and Dong-Sung Cho. 2000. *Tiger Technology: The Creation of a Semiconductor Industry in East Asia*. Cambridge, UK: Cambridge University Press.

Mazzarol, Tim and Geoffrey N. Soutar. 2002. "'Push-Pull' Factors Influencing International Student Destination Choice." *The International Journal of Educational Management* 16 (2/3):82–90.

Mazzucato, Mariana. 2013. *The Entrepreneurial State: Debunking Public vs. Private Sector Myths*. London: Anthem Press.
McDowell, Gary S. et al. 2015. "Shaping the Future of Research: A Perspective from Junior Scientists." *F1000 Research* 3:291. doi: 10.12688/f1000research.5878.2.
McDowell, Linda. 2017. "Youth, Children, and Families in Austere Times: Change, Politics and a New Gender Contract." *Area* 49(3):311–316.
McGrayne, Sharon B. 1998. *Nobel Prize Women in Science: Their Lives, Struggles, and Momentous Discoveries*. Washington, DC: John Henry Press.
Melin, Göran. 2004. "Postdoc Abroad: Inherited Scientific Contacts or Establishment of New Networks?" *Research Evaluation* 13(2):95–102.
Melin, Göran and Kerstin Janson. 2006. "What Skills and Knowledge Should a PhD Have? Changing Preconditions for PhD Education and Post Doc Work." In *The Formative Years of Scholars*, edited by U. Teichler, 105–118. Wenner Gren International Series 83. London: Portland Press.
Merton, Robert K. 1942. "Science and Technology in a Democratic Order." *Journal of Legal and Political Sociology* 1:115–126.
Merton, Robert K. 1968. "The Matthew Effect in Science: The Reward and Communication Systems of Science Are Considered." *Science* 159(3810):56–63.
Merton, Robert K. 1973. *The Sociology of Science: Theoretical and Empirical Investigations*. Chicago, IL: The University of Chicago Press.
Milio, Simona, Riccardo Lattanzi, Francesca Casadio et al. 2012. *Brain Drain, Brain Exchange and Brain Circulation: The Case of Italy Viewed from a Global Perspective*. Report for the Aspen Institute Italia. Rome, Italy: Aspen Institute Italia.
Ministry of Science and Technology (MOST). 2017. *National Science and Technology Development Plan (2017–2020)*. Taipei: Ministry of Science and Technology, R.O.C. www.most.gov.tw/most/attachments/2abb3ec5-78f0-4c00-80c4-063294ec76ab.
Ministry of Trade and Industry. 2006. *Sustaining Innovation-Driven Growth: Science and Technology 2010 Plan*. Singapore: Ministry of Trade and Industry.
Ministry of Trade and Industry. 2011. *Research Innovation Enterprise 2015: Singapore's Future*. Singapore: Research Innovation and Enterprise Secretariat.
Mitchell, Jeremy S. et al. 2013. *The 2013 Canadian Postdoc Survey: Painting a Picture of Canadian Postdoctoral Scholars*. Canadian Association of Postdoctoral Scholars and Mitacs.

Moguérou, Philippe. 2005. "Doctoral and Postdoctoral Education in Science and Engineering: Europe in the International Competition." *European Journal of Education Research, Development and Policy* 40(4):367–392.

Monosson, Emily. 2011. *Motherhood, the Elephant in the Laboratory: Women Scientists Speak Out*. Ithaca, NY: Cornell University Press.

Moss-Racusin, Corinne A., John F. Dovidio, Victoria L. Brescoll et al. 2012. "Science Faculty's Subtle Gender Biases Favor Male Students." *Proceedings of the National Academy of Sciences* 109(41):16474–16479.

Murakami, Yukiko. 2014. "Incentives for International Migration of Scientists and Engineers to Japan." *International Migration* 47(4):67–91.

Natarajan, Mangai. 2016. "Rapid Assessment of 'Eve Teasing' (Sexual Harassment) of Young Women during the Commute to College in India." *Crime Science* 5(6). doi: 10.1186/s40163-016-0054-9.

National Institute of Science, Technology and Development Studies (NISTADS). 2009. *India Science & Technology 2008*. New Delhi, India: NISTADS.

National Research Council. 2000. *Addressing the Nation's Changing Needs for Biomedical and Behavioral Scientists*. Washington, DC: The National Academies Press.

National Research Council. 2009. *A New Biology for the 21st Century*. Washington, DC: National Academies Press.

National Research Council. 2014. *The Postdoctoral Experience Revisited*. Washington, DC: National Academies Press.

National Science Board, National Science Foundation. 2020. *Science and Engineering Indicators 2020: The State of U.S. Science and Engineering 2020*. NSB-2020-1. Alexandria, VA. Accessed on January 12, 2021. https://ncses.nsf.gov/pubs/nsb20201/.

National Science Foundation. 2006. *U.S. Doctorates in the 20th Century*. NSF Special Report, 06-319. Arlington, VA: National Science Foundation. Accessed on June 24, 2021. https://wayback.archiveit.org/5902/20150817174927/http://www.nsf.gov/statistics/nsf06319/.

National Science Foundation. 2018. "Doctorate Recipients from U.S. Universities, 2016," table 14. Science and Engineering Doctorates. National Science Foundation. www.nsf.gov/statistics/2018/nsf18304/data/tab14.pdf.

Nature Materials. 2009. "Challenges for Science in India." *Nature Materials* 8:361.

Nature Publishing Group. 2015. *Turning Point: Chinese Science in Transition*. Berlin, Germany: Springer Nature.

Nature. 2007. "Asia on the Rise." *Nature* 447:885.

Nature. 2008. "One Woman Is Still Not Enough." *Nature* 451:865.

Nature. 2015a. "A Nation with Ambition." *Nature* 521:125.
Nature. 2015b. "Research Management: Priorities for Science in India." *Nature* 521:151–155.
Nature. 2016. "Catch Them If You Can." *Nature* 535:S68–S76.
Needham, Joseph. 1954–2005. *Science and Civilization in Asia.* Volumes 1–6.3. Cambridge, UK: Cambridge University Press.
Ng, Pak Tee. 2013. "The Global War for Talent: Responses and Challenges in the Singapore Higher Education System." *Journal of Higher Education Policy and Management* 35(3):280–292.
Ng, Pak Tee and Charlene Tan. 2010. "The Singapore Global Schoolhouse: An Analysis of the Development of the Tertiary Education Landscape in Singapore." *International Journal of Educational Management* 24(3):178–188.
Nisbett, Richard E. 2003. *The Geography of Thought: How Asians and Westerners Think Differently and Why.* London: Hachette.
Nowotny, Helga. 2017. *An Orderly Mess.* Budapest, Hungary: Central European University Press.
Nowotny, Helga, Peter Scott and Michael Gibbons. 2001. *Re-thinking Science: Knowledge and the Public in an Age of Uncertainty.* London: Polity Press.
Nuffield Council on Bioethics. 2014. *The Culture of Scientific Research in the UK.* London: Nuffield Council on Bioethics.
Oberg, Kalervo. 1960. "Cultural Shock: Adjustment to Cultural Environments." *Practical Anthropology* 7(4):177–182.
Oishi, Nana. 2005. *Women in Motion: Globalization, State Policies, and Labor Migration in Asia.* Stanford, CA: Stanford University Press.
Oishi, Nana. 2012. "The Limits of Immigration Policies: The Challenges of Highly Skilled Migration in Japan." *American Behavioral Scientist* 56(8):1080–1100.
Ong, Aihwa. 1999. *Flexible Citizenship: The Cultural Logics of Transnationality.* Durham, NC: Duke University Press.
Ong, Aihwa. 2016. *Fungible Life: Experiment in the Asian City of Life.* Durham, NC: Duke University Press.
Ortiga, Yasmin Y., Meng-Hsuan Chou, Gunjan Sondhi and Jue Wang. 2018. "Academic 'Centres,' Epistemic Differences and Brain Circulation." *International Migration* 56(5):90–105.
Ortner, Sherry. 2003. "East Brain, West Brain." *New York Times*, April 20. Accessed on March 2, 2021. www.nytimes.com/2003/04/20/books/east-brain-west-brain.html.
Parayll, Govindan. 2005. "From 'Silicon Island' to 'Biopolis of Asia': Innovation Policy and Shifting Competitive Strategy in Singapore." *California Management Review* 47(2):50–73.

Paul, Anju M. 2011. "Stepwise International Migration: A Multistage Migration Pattern for the Aspiring Migrant." *American Journal of Sociology* 116(6):1842–1886.

Paul, Anju M. 2015. "Capital and Mobility in the Stepwise International Migrations of Filipino Migrant Domestic Workers." *Migration Studies* 3(3):438–459.

Paul, Anju M. 2017. *Multinational Maids: Stepwise Migration in a Global Labor Market*. Cambridge, UK: Cambridge University Press.

Paul, Anju M. 2018. "The Destination Decision of Asian Postdoctoral Trainees: Advice from Asian-Born, Western-Trained Bioscientists." In *High-Skilled Migration: Drivers, Dynamics and Policies*, edited by M. Czaika, 279–300. Oxford: Oxford University Press.

Paul, Anju M. and Victoria Long. 2016. "Human-Capital Strategies to Build World-Class Research Universities in Asia: Impact on Global Flows." In *The Transnational Politics of Higher Education: Contesting the Global / Transforming the Local*, edited by M.-H. Chou, I. A. Kamola and T. Pietsch, 130–155. London: Routledge.

Paul, Anju M. and Victoria Long. 2017. "Where to Train: Shifts in the Doctoral Destination Advice Given to Asian Bioscience Students." *Global Studies* 10(3):1–18.

Paul, Anju M. and Brenda S. A. Yeoh. 2020. "Studying Multinational Migrations, Speaking Back to Migration Theory." *Global Networks* 21(1):3–17.

Pellens, Maikel. 2012. "The Motivations of Scientists as Drivers of International Mobility Decisions." Working paper OR 1202, Faculty of Business and Economics, Katholieke Universiteit Leuven. https://lirias.kuleuven.be/bitstream/123456789/337077/1/MSI1202.pdf.

Pessar, Patricia R. 1996. "Women, Gender, and International Migration across and beyond the Americas: Inequalities and Limited Empowerment." Expert Group Meeting on International Migration and Development in Latin America and the Caribbean, November 30–December 2, 2005. Mexico City, Mexico: UNDP.

Peters, Michael A. and Tina Besley. 2018. "China's Double First-Class University Strategy." *Educational Philosophy and Theory* 50 (12):1075–1079.

Pfau-Effinger, Birgit. 1994. "The Gender Contract and Part-Time Paid Work by Women: Finland and Germany Compared." *Environment and Planning A: Economy and Space* 26(9):1355–1376.

Pietsch, Tamson. 2013. *Empire of Scholars: Universities, Networks and the British Academic World, 1850–1939*. Manchester, UK: Manchester University Press.

Portes, Alejandro and Rubén Rumbaut. 2001. *Legacies: The Story of the Immigrant Second Generation.* Berkeley, CA: University of California Press.

Portes, Alejandro and Min Zhou. 1993. "The New Second Generation: Segmented Assimilation and Its Variants." *The ANNALS of the American Academy of Political and Social Science* 530(1):74–96.

Prakash, Gyan. 1999. *Another Reason: Science and the Imagination of Modern India.* Princeton, NJ: Princeton University Press.

Purkayastha, Bandana. 2005. "Skilled Migration and Cumulative Disadvantage: The Case of Highly Qualified Asian Indian Immigrant Women in the US." *Geoforum* 36(2):181–196.

Quah, Stella R. 2009. *Families in Asia: Home and Kin.* New York: Routledge.

Raghuram, Parvati. 2006. "Gendering Medical Migration: Asian Women Doctors in the UK." In *Migrant Women and Work*, edited by A. Agrawal, 73–94. New Delhi: Sage Publications India.

Raghuram, Parvati. 2013. "Theorising the Spaces of Student Migration." *Population, Space and Place* 19(2):138–154.

Ramirez, Francisco O. 2010. "Accounting for Excellence: Transforming Universities into Organizational Actors." In *Higher Education, Policy, and the Global Competition Phenomenon*, edited by L. M. Portnoi, V. D. Rust and S. S. Bagley, 43–58. London: Springer.

Rapoport, Hillel, Sulin Sardoschau and Arthur Silve. 2020. "Migration and Cultural Change." CESifo working paper no. 8547. Ludwig-Maximilians-Universitaet and the Ifo Institute. Munich, Germany. Accessed on March 14, 2021. https://ssrn.com/abstract=3689469.

Ray, Debraj. 2006. "Aspirations, Poverty, and Economic Change." In *Understanding Poverty*, edited by A. V. Banerjee, R. Bénabou and D. Mookherjee, 409–421. Oxford: Oxford University Press.

Reckwitz, Andreas. 2002. "Toward a Theory of Social Practices." *European Journal of Sociology* 5(2):243–263.

Regets, Mark C. 1998. "Has the Use of Postdocs Changed?" Issue Brief, Division of Science Resources Studies, NSF 99-310. Washington, DC: National Science Foundation.

Reuben, Ernesto, Paola Sapienza and Luigi Zingales. 2014. "How Stereotypes Impair Women's Careers in Science." *Proceedings of the National Academy of Sciences of the United States of America* 111(12):4403–4408.

Rinne, Tiffany, G. Daniel Steel and John Fairweather. 2012. "Hofstede and Shane Revisited: The Role of Power Distance and Individualism in National-Level Innovation Success." *Journal of Cross-Cultural Research* 46(2):91–108.

Rinne, Tiffany, G. Daniel Steel and John Fairweather. 2013. "The Role of Hofstede's Individualism in National-Level Creativity." *Creativity Research Journal* 25(1):129–136.
Robertson, Susan L. 2006. "Brain Drain, Brain Gain and Brain Circulation." *Globalisation, Societies and Education* 4(1):1–5.
Rosenzweig, Mark R. 2007. "Higher Education and International Migration in Asia: Brain Circulation." Paper presented at the Regional Bank Conference on Development Economics, Higher Education and Development, Beijing, January 16–17.
Roy, Denny. 2003. *Taiwan: A Political History*. Ithaca, NY: Cornell University Press.
Sandberg, Sheryl and Nell Scovell. 2013. *Lean In: Women, Work, and the Will to Lead*. New York: Alfred A. Knopf.
Sands, Aimée. 2009. "Never Meant to Survive, a Black Woman's Journey: An Interview with Evelyn Hammond." In *Women, Science, and Technology*, edited by M. Wyer, M. Barbercheck, D. Giesman et al., 31–39. New York: Routledge.
Sarukkai, Sundar. 2015. "Challenges for STEM Education in India." In *International Science and Technology Education: Exploring Culture, Economy and Social Perceptions*, edited by O. Renn, N. C. Karafyllis, A. Hohlt et al., 69–84. New York: Routledge.
Saxenian, AnnaLee. 2002. "Transnational Communities and the Evolution of Global Production Networks: The Cases of Taiwan, China and India." *Industry and Innovation* 9(3):183–202.
Saxenian, AnnaLee. 2005. "From Brain Drain to Brain Circulation: Transnational Communities and Regional Upgrading in India and China." *Studies in Comparative International Development* 40(2):35–61.
Schiebinger, Londa. 1999. *Has Feminism Changed Science?* Cambridge, MA: Harvard University Press.
Science, Technology and Innovation Center (STIC). 2004. *Knowledge and Innovation 2004 Report*. Taipei, Taiwan: STIC.
Scientific Advisory Council to the Prime Minister. 2011. *Bi-Annual Report*. New Delhi, India: Ministry of Science & Technology.
Scoones, Ian. 2002. "Biotech Science, Biotech Business: Current Challenges and Future Prospects." *Economic and Political Weekly* 7(27):2725–2733.
Shane, Scott. 1993. "Cultural Influences on National Rates of Innovation." *Journal of Business Venturing* 8(1):59–73.
Shapin, Steven and Simon Schaffer. 1985. *Leviathan and the Air-Pump: Hobbes, Boyle, and the Experimental Life*. Princeton, NJ: Princeton University Press.
Sharma, Yojana. 2021. "New and Strengthened Research Priorities Post-Pandemic." *University World News*. March 6. Accessed on March 26,

2021. www.universityworldnews.com/post.php?story=20210304105937958.
Slaughter, Anne-Marie. 2012. "Why Women Still Can't Have It All." *The Atlantic*, July/August. Accessed on March 20, 2021, www.theatlantic.com/magazine/archive/2012/07/why-women-still-cant-have-it-all/309020/.
Smaglik, Paul. 2003. "Filling Biopolis Singapore." *Nature* 425:746–747.
Smith, Heather A. and James D. McKeen. 2011. "Instilling a Knowledge-Sharing Culture." Working paper, Queens University School of Business, Kingston, Ontario. Accessed on July 15, 2021. https://citeseerx.ist.psu.edu/viewdoc/download?doi=10.1.1.199.4584&rep=rep1&type=pdf.
Stark, Oded and David E. Bloom. 1985. "The New Economics of Labor Migration." *American Economic Review* 75(2):173–178.
Stark, Oded and J. E. Taylor. 1989. "Relative Deprivation and International Migration." *Demography* 26(1):1–14.
Stephan, Paula, Chiara Franzoni and Giuseppe Scellato. 2013. "Choice of Country by the Foreign Born for PhD and Postdoctoral Study: A Sixteen-Country Perspective." National Bureau of Economic Research Working Paper Series, no. 18809, February. Accessed on July 15, 2021. www.nber.org/papers/w18809.pdf.
Stephan, Paula E. and Jennifer Ma. 2005. "The Increased Frequency and Duration of the Postdoctorate Career Stage." *American Economic Review* 95(2):71–75.
Stolte-Heiskanen, Veronica. 1991. *Women in Science: Token Women or Gender Equality?* New York: Berg.
Stone, Richard. 2008. "South Korea Aims to Boost Status as Science and Technology Powerhouse." *Science*, December 23. Accessed on March 13, 2021. www.sciencemag.org/news/2008/12/south-korea-aims-boost-status-science-and-technology-powerhouse.
Stone, Richard. 2012. "India Rising." *Science* 335(6071):904–910.
Storer, Norman W. 1966. *The Social System of Science*. New York: Holt, Reinhart and Winston.
Subbarayappa, B. V. 2013. *Science in India: A Historical Perspective*. New Delhi, India: Ruha Publications.
Subrahmanyan, Lalita. 1998. *Women Scientists in the Third World: The Indian Experience*. Thousand Oaks, CA: Sage Publications.
Subramanian, Ajantha. 2015. "Making Merit: The Indian Institutes of Technology and the Social Life of Caste." *Comparative Studies in Society and History* 57(2):291–322.
Szelényi, Katalin. 2006. "Students without Borders? Migration Decision-Making among International Graduate Students in the U.S." *Knowledge, Technology & Policy* 19(3):64–86.

Taiwan Today. 2018. "Tsai Inaugurates National Biotechnology Research Park in Taipei." Taiwan Today, Ministry of Foreign Affairs, Republic of China, October 16. Accessed on January 12, 2021. https://taiwantoday.tw/news.php?unit=2,6,10,15,18&post=143450.

Tambyah, Paul A. 2005. "Selection of Medical Students in Singapore: A Historical Perspective." *Annals of the Academy of Medicine Singapore* 34(6):147C–151C.

Tan, B. T. G. 2017. "The Science Council and Singapore Science in the '60s and '70s." In *50 Years of Science in Singapore*, edited by B. T. G. Tan, Hock Lim and K. K. Phua, 1–22. Singapore: World Scientific.

Thaler, Richard H. and Cass R. Sunstein. 2008. *Nudge: Improving Decisions about Health, Wealth, and Happiness*. New Haven, CT: Yale University Press.

Thampuran, Raj et al. 2017. "Science, Technology and Open Innovation – The A*STAR Journey." In *50 Years of Science in Singapore*, edited by B. T. G. Tan, H. Lim and K. K. Phua, 39–50. Singapore: World Scientific.

Thaxton, Jr., Ralph A. 2008. *Catastrophe and Contention in Rural China: Mao's Great Leap Forward Famine and the Origins of Righteous Resistance in Da Fo Village*. Cambridge, UK: Cambridge University Press.

Thomas, Renny. 2020. "Brahmins as Scientists and Science as Brahmins' Calling: Caste in an Indian Scientific Research Institute." *Public Understanding of Science* 29(3):306–318.

Thurs, Daniel P. 2007. *Science Talk: Changing Notions of Science in American Culture*. Camden, NJ: Rutgers University Press.

Traweek, Sharon. 1992. *Beamtimes and Lifetimes: The World of High Energy Physicists*. Harvard, MA: Harvard University Press.

Traweek, Sharon. 1993. "An Introduction to Cultural and Social Studies of Sciences and Technologies." *Culture, Medicine and Psychiatry* 17:3–25.

Tsai, Chin-Chung and Pi-Chu Kuo. 2008. "Cram School Students' Conceptions of Learning and Learning Science in Taiwan." *International Journal of Science Education* 30(3):353–375.

Turpin, Tim, Richard Woolley and Jane Marceau. 2010. "Scientists across the Boundaries: National and Global Dimensions of Scientific and Technical Human Capital (STHC) and Policy Implications for Australia." *Asian and Pacific Migration Journal* 19(1):65–86.

Tzeng, Rueyling. 2006. "Reverse Brain Drain: Government Policy and Corporate Strategies for Global Talent Searches in Taiwan." *Asian Population Studies* 2(3):239–256.

UNESCO. 2017. *Cracking the Code: Girls' and Women's Education in Science, Technology, Engineering and Mathematics (STEM)*. Paris, France: UNESCO.

Upadhya, Carol. 2013. "Return of the Global Indian: Software Professionals and the Worlding of Bangalore." In *Return: Nationalizing Transnational Mobility in Asia*, edited by B. Xiang, B. S. A. Yeoh and M. Toyota, 141–161. Durham, NC: Duke University Press.

Vale, Ron. 2014. "Is India Ready to Boost Its Postdoctoral Training?" IndiaBioscience, May 8. Accessed on June 25, 2021. https://indiabioscience.org/columns/indian-scenario/is-india-ready-to-boost-its-post-doctoral-training.

Vale, Ronald D. and Karen Dell. 2009. "The Biological Sciences in India." *Journal of Cell Biology* 184(3):342–353.

Van Brunt, Jennifer. 1988. "In China, the Dragon Searches for the Pearl." *Nature Biotechnology* 6(1):863–864.

Van Noorden, Richard. 2015. "India by the Numbers." *Nature* 521:142–143.

Varrel, Aurélie. 2011. "Gender and Intergenerational Issues in the Circulation of Highly Skilled Migrants: The Case of Indian IT Professionals." In *Gender, Generations and the Family in International Migration*, edited by A. Kraler, E. Kofman, M. Kohli et al., 337–353. Amsterdam, the Netherlands: Amsterdam University Press.

Velema, Thijs A. 2012. "The Contingent Nature of Brain Gain and Brain Circulation: Their Foreign Context and the Impact of Return Scientists on the Scientific Community in Their Country of Origin." *Scientometrics* 93:893–913.

Vertovec, Steven. 2003. "Migration and Other Modes of Transnationalism: Towards Conceptual Cross-fertilization." *International Migration Review* 37(3):641–665.

Veugelers, Reinhilde. 2012. "The Rise of Asia in Science." *Bruegel*, October 25. Accessed on January 27, 2021. www.bruegel.org/2012/10/the-rise-of-asia-in-science/.

Veugelers, Reinhilde. 2017. "The Challenge of China's Rise as a Science and Technology Powerhouse." Bruegel Policy Contribution, no. 19, July 5. Accessed on March 13, 2021. www.bruegel.org/2017/07/the-challenge-of-chinas-rise-as-a-science-and-technology-powerhouse/.

Visvanathan, Shiv and Chandrika Parmar. 2002. "A Biotechnology Story: Notes from India." *Economic and Political Weekly* 37(27):2714–2724.

Vogel, Ezra F. 2013. *Deng Xiaoping and the Transformation of China*. Cambridge, MA: Harvard University Press.

Wagner, Caroline. 2021. "Scrutiny of Chinese Researchers Threatens Innovation." *University World News*, January 30. Accessed on June 19, 2021. www.universityworldnews.com/post.php?story=20210128141727753.

Wang, Cheng-Lung and Pey-Yan Liou. 2017. "Students' Motivational Beliefs in Science Learning, School Motivational Contexts, and Science

Achievement in Taiwan." *International Journal of Science Education* 39(7):898–917.
Wang, Hsiu-Yun and Joel F. Stocker. 2010. "Women Scientists in Taiwan: An Update." *The Kaohsiung Journal of Medical Sciences* 26 (6):S35–SS40.
Wang, Qingfang, Li Tang and Huiping Li. 2014. "Return Migration of the Highly Skilled in Higher Education Institutions: A Chinese University Case." *Population, Space and Place* 21(8):771–787.
Waters, Johanna and Rachel Brooks. 2010. "Accidental Achievers? International Higher Education and Class Reproduction and Privilege in the Experiences of UK Students Overseas." *British Journal of Sociology of Education* 31(2):217–228.
Watson, James D. 1968. *The Double Helix: A Personal Account of the Discovery of the Structure of DNA*. New York: Simon & Schuster.
Watts, Ruth. 2007. *Women in Science: A Social and Cultural History*. London: Routledge.
Weber, Max. 1958. "Science as a Vocation." In *From Max Weber: Essays in Sociology*, 129–156. Translated by H. H. Gerth and C. Wright Mills. New York: Oxford University Press.
Welch, Anthony and Jie Hao. 2016. "Global Argonauts: Returnees and Diaspora as Sources of Innovation in China and Israel." *Globalisation, Societies and Education* 14(2):272–297.
Widén, Gunilla and Preben Hansen. 2012. "Managing Collaborative Information Sharing: Bridging Research on Information Culture and Collaborative Information Behaviour." *Information Research* 17(4):538. Accessed on March 14, 2021. http://informationr.net/ir/17-4/paper538.html#.YE1Kw5MzZTY.
Willsher, Kim. 2021. "French Self-Esteem Hit after Pasteur Institute Abandons Covid Vaccine." *Guardian*, January 26. Accessed on February 13, 2021. www.theguardian.com/world/2021/jan/26/french-self-esteem-hit-after-pasteur-institute-abandons-covid-vaccine.
Woetzel, Jonathan and Jeongmin Seong. 2020. "What Is Driving Asia's Technological Rise?" *Japan Times*, December 24. Accessed on January 27, 2021. www.japantimes.co.jp/opinion/2020/12/24/commentary/world-commentary/asia-technological-rise/.
Wong, Grace. 2008. "Can China's Supply of Scientific Talent Keep up with Demand." *Nature Biotechnology* 26(2):243–244.
Wong, Joseph. 2005. "Re-making the Developmental State in Taiwan: The Challenges of Biotechnology." *International Political Science Review* 26(2):169–191.
Wong, Joseph. 2011. *Betting on Biotech: Innovation and the Limits of Asia's Developmental State*. Ithaca, NY: Cornell University Press.

Wong, Ting-Hong. 2000. "State Formation, Hegemony, and Nanyang University in Singapore, 1953 to 1965." *Formosan Education and Society* 1(1):59–85.
Woolley, Richard, Tim Turpin, Jane Marceau et al. 2008. "Mobility Matters: Research Training and Network Building in Science." *Comparative Technology Transfer and Society* 6(3):159–186.
Xi, Qiaojuan and Zhang Aixiu. 2010. *China's Science, Technology and Education*. Beijing, China: China Intercontinental Press.
Xia, Xiahui. 2014. "Biotechnology Industry in China." Masters thesis, University of Massachusetts Lowell.
Xiang, Biao. 2007. *Global 'Body Shopping': An Indian Labor System in the Information Technology Industry*. Princeton, NJ: Princeton University Press.
Xiang, Biao. 2013. "Return and the Reordering of Transnational Mobility in Asia." In *Return: Nationalizing Transnational Mobility in Asia*, edited by B. Xiang, B. S. A. Yeoh and M. Toyota, 1–20. Durham, NC: Duke University Press.
Xiang, Biao and Wei Shen. 2009. "International Student Migration and Social Stratification in China." *International Journal of Educational Development* 29(5):513–522.
Xie, Yu and Alexandra A. Killewald. 2012. *Is American Science in Decline?* Harvard, MA: Harvard University Press.
Xie, Yu and Kimberlee A. Shauman. 2003. *Women in Science: Career Processes and Outcomes*. Cambridge, MA: Harvard University Press.
Xie, Yu, Chunni Zhang and Qing Lai. 2014. "China's Rise as a Major Contributor to Science and Technology." *Proceedings of the National Academy of Sciences of the United States of America* 111(26):9437–9442.
Yeoh, Brenda S. A. and Lai Ah Eng. 2008. "'Talent Migration in and out of Asia: Challenges for Policies and Places." *Asian Population Studies* 4(3):235–245.
Yeoh, Francis. 2017. "R&D in Singapore – The Early Years of NSTB." In *50 Years of Science in Singapore*, edited by B. T. G. Tan, H. Lim and K. K. Phua, 23–38. Singapore: World Scientific.
Yeoh, Keat-Chuan. 2008. "Singapore's Biomedical Sciences Landscape." *Journal of Commercial Biotechnology* 14(2):141–148.
Zeithammer, Robert and Ryan P. Kellogg. 2013. "The Hesitant Hai Gui: Return-Migration Preferences of U.S.-Educated Chinese Scientists and Engineers." *Journal of Marketing Research* 50(5):644–663.
Zeng, Wei. 2008. "China's Challenge." *Nature* 452:1028–1029.
Zhang, Fangzhu, Philip Cooke and Fulong Wu. 2010. "State-Sponsored Research and Development: A Case Study of China's Biotechnology." *Regional Studies* 45(5):575–595.

Zhang, Liang, Liang Sun and Wei Bao. 2017. "The Rise of Higher Education and Science in China." *International Perspectives on Education and Society* 33:141–172.

Zhang, Shuxian, Zezhou Wang, Ruijie Chang et al. 2020. "COVID-19 Containment: China Provides Important Lessons for Global Response." *Frontiers of Medicine* 14(2):215–219.

Zhou, Min. 1997. "Segmented Assimilation: Issues, Controversies, and Recent Research on the New Second Generation." *International Migration Review* 31(4):975–1008.

Zhou, Yingying. 2015. "The Rapid Rise of a Research Nation." *Nature* 528: S170–173.

Zhou, Yuefang, Divya Jindal-Snape, Keith Topping et al. 2008. "Theoretical Models of Culture Shock and Adaptation in International Students in Higher Education." *Studies in Higher Education* 33(1):63–75.

Zijlstra, Judith. 2020. "Stepwise Migration of Iranian Students from Turkey to the West." *Geographical Research* 58(4):403–415.

Zucker, Lynne G. and Michael R. Darby. 1996. "Star Scientists and Institutional Transformation: Patterns of Invention and Innovation in the Formation of the Biotechnology Industry." *Proceedings of the National Academy of Sciences* 93:12709–12716.

Zumeta, William. 1985. *Extending the Educational Ladder: The Changing Roles of Postdoctoral Education in the United States.* Springfield, VA: National Technical Information Service.

Zweig, David and Changgui Chen. 1995. *China's Brain Drain to the United States: Views of Overseas Chinese Students and Scholars in the 1990s.* China Research Monograph 47. Berkeley: Institute of East Asian Studies, University of California.

Zweig, David, Changgui Chen and Stanley Rosen. 2004. "Globalization and Transnational Human Capital: Overseas and Returnee Scholars to China." *The China Quarterly* 179 (September):735–757.

Index

863 Program, 48

A*STAR, 65, 67, 68, 177, 252, 293
Academia Sinica, 4, 30, 74, 75, 76, 109, 172, 229, 233
Afghanistan, 298
Altbach, Philip, 54
Ambition, research, 21, 181
  and return migration, 132, 134, 151–152, 168–174
  and westward migration, 113
  and women, 187–188, 205, 206, 214
  in China, 133
  in India, 131
  in the USA, 131, 238, 256
  lack of, 276
  of returned scientists, 287
  raising, 40, 259, 269, 274–279
  scope of, 22, 39, 121, 130–134
Appadurai, Arjun, 96, 130
Applied research, 4, 66, 69–70, 231, 294-6
Arranged marriage
  and migration, 200
Asia
  definition of, 26
Asian science
  emergence of, 300–302, 304
  rejecting the concept, 7, 118, 304
Asian scientists
  definition of, 26
  next generation of, 172, 203, 286
Asian Tigers, 24
Assimilation
  challenges of, 159
  cultural, 153, 260
  definition of, 159
  upon return, 154
Assimilation, segmented, 223

ATREE, xxiii
Australia
  as destination choice, 56
  as origin country, 89
  as Western, 26
Autonomy, degree of, 22, 121, 134–137

Bangalore, xviii, 55, 59, 122, 243, 244, 267, 270
  and ecology, 243
Basalla, George, 57, 58
Basic research, 4, 48, 50, 60, 66, 69–70, 231, 233, 296
Bharat Biotech, 60
Bhutan, 299
Biological sciences
  and women, 187
  definition of, 11
  in China, 48
  interest in, 196
Biopolis, 3, 67–68, 240
Biotechnology industry
  in China, 49–50
  in India, 62
  in Singapore, 67, 71
  in Taiwan, 76, 77
Boden, Margaret, 127
Boundary-work, 314
Bourdieu, Pierre, 9, 10, 11, 59, 120
Brain boomerang, 45, *See also* Brain circulation
  strategy, 57
Brain circulation, 17–19, 35, 37, 47, 48, 62, 89, 288
  delays in, 47
  from Europe to the United States, 89
  historical, 48
  lateral, 89, 95
  shifts in, 95
Brain drain, xiv, 17, 45, 61, 198

333

Brain gain, 17, 64–65
  strategy, 56, 281, 298
Brazil, 61
Brenner, Sydney, 66
Bulletin Board System, 98
Bureaucracy, government
  and corruption, 238
  in China, 238
  in India, xvi
  in Singapore, 229, 230, 231
  power of, 231
Bureaucracy, university
  absence of, 129
  in Asia, 235–240
  in India, 79
Bureaucrats
  versus scientists, 230

C9 group, 44
Cambridge University, 28, 29, 116
  reputation of, 108, 274, 276
  scientific culture in, 129
Canada
  as destination choice, 4, 26, 56, 91, 116, 245
Capital
  accumulation of, 11
  and gender, 218
  and status, 9
  constraints, xii, 101, 105
  cultural, 9
  definition of, 9
  economic, 9, 260
  human, 9, 50, 56, 198
  mobility of, 79
  scientific, 9
  scientific cultural, 89, 104, 113, 114, 176, 251
  scientific human, 89, 175, 180, 288
  scientific social, 96, 104, 113, 114
  social, 9, 260
  symbolic, 113
  types of, 9
  Western-acquired, 175
Career course, 152, *See also* Life course
  and life course, 167, 184, 185, 207, 216, 217, 218
  and migration, 113, 210
  and mobility expectations, 149
  and return migration, 177
  for academic scientists, 12
  for women scientists, 216, 291
  in academia, 14–15
  variations in, 15–16
Cassarino, Jean-Pierre, 267
Cell, 10
Centre for Cellular and Molecular Biology, 59, 168
Changing demographics
  in China, 265, 266
  in India, 24, 56
  in Singapore, 63, 265
  in Taiwan, 264, 284
Children
  and return migration, 37, 165
  and women scientists, 190
China
  as origin, 265
  corruption in, 229
  higher education in, 48
  history of, 42–43, 47–48
China-Taiwan relations, 77
Chinese Academy of Sciences, xviii, xxi, 30, 46, 51, 241
Chinese economy
  development of, 46
  growth of, 43
  liberalization of, 43, 155
  size of, 43
Chinese-US Biochemistry Examination and Application, 92, 93, *See also* Doctoral scholarships
Choi, Susanne, 193
Class privilege
  and gender, 194
Clinical research
  in Singapore, 69
  in the United Kingdom, 123
Coleman, Alan, 68
Colonial science, 57
Columbia University, 28, 29
Communality, in science, 21, 22, 119
  violation of, 22
Communication, approach to, 122, 123, 129
  changing, 278
  external, 144
  in the UK, 129
  internal, 143
Communist Party of China, 23, 42

Index 335

Confucian culture, 165, 277
Contexts of reception, 223
Contexts of return
  scientific, 223
Copeland, Nancy, 68
Copeland, Neal, 68
Cornell University, xxi, 28
Cost of migration, 93, 99–100
Council of Scientific and Industrial Research, 59
Covid-19, 52, 53, 60
Creativity
  and individualism, 135
  and isolation, 148
  and scientific research, 268, 275
  culture of, 121
  in China, 128, 297
  in India, 276
  in Singapore, 128
  in the United States, 124, 130
  in the West, 127
  types of, 127
Critical thinking, 118, 143, 145, *See also* Scientific knowledge, attitude towards, Organized skepticism, in science
  absence of, 125, 126
  in Taiwan, 124
  in Western universities, 123
Cultural change, 23
  in Asia, 279
  in India, 285
  slow pace of, 285
Cultural Revolution, 43, 48, 83
  impact of, 23, 43, 115
Culture of migration, 94–96
  to the West, 90, 94
Culture shock, 167, 184, *See also* Gender shock
  reverse, 208, 277
  scientific, 124, 137, 184
Cumulative disadvantage, 217
Curie, Marie, 203
Curiosity
  culture of, 124, 129, 259, 269

Darwin, Charles, 147
Debates, scientific
  importance of, 277
Deng, Xiaoping, 23, 43, 44, 45, 48

Destination decision, 87, 105–108
Destination hierarchy, 87, 105
Dewey, John, 123
Diaspora
  Chinese, 50
Difference, attitudes towards, 22, 122
  broadening, 279–283
Diffusion, cultural, 23, 259, 261, 262, 290
Diffusion, informational, 260, 262, 290
Diffusion, reputational, 260, 262, 290
Discrimination
  absence of, 159
  in Europe, 160
  in the sciences, 22, 299
  in the United States, 160
Discrimination, anti-Asian, 160, 161, 248, 249
  absence of, 160
  compared across countries, 160
Discrimination, anti-Muslim, 204
Discrimination, gender, 184, 197, 204, 207, *See* Gender discrimination
Discrimination, organizational, 16, 212, 248
Discrimination, racial, 160, 208
Disinterestedness, in science, 21, 58, 119
  violation of, 21
Dissimilation, xvi, 154, 263, *See also* Assimilation
Diversity
  as a goal, 140, 180
  in Asian science, 142
  in European science, 141, 161
  in science, 140, 145, 163, 252, 274, 279, 283
  in Taiwan, 280
  value of, 163, 274
Divorce, 205, 206, 209, 210
  among Asian women scientists, 205
  and Asian women scientists, 206, 208, 209, 210, 217, 218
Doctoral scholarships, 98
  and destination decision, 108
  and women, 199
  from Asia, 92
  from China, 199
  from Singapore, 93, 176
  from Taiwan, 93, 108
  government, 19, 99, 157
  in India, 92, 196

Doctoral scholarships (cont.)
  to Asia, 109, 110
  to Singapore, 109
  to Taiwan, 109
  to the US, 92
Doctoral students
  quality of, 93, 176, 225, 249
Domestic workers, 194
  and Asian women scientists, 190, 218
  and stepwise migration, xii, 105
Double First Class University Plan, 44
du Sautoy, Marcus, 127
Duke University, 64, 68
Duke-NUS Medical School, 64, 68

Education reform
  in China, 266
Education, overseas
  as investment, 46, 56
Education, science. *See* Science
  Education
Elman, Benjamin, 48
Emigration policies
  in China, 45
Engineering education, 53
English language
  and science, 280
Entrepreneurial state, 78
Eve teasing, 212
Examinations, national
  in China, 43, 44
  in India, 190
Exchange programs, student, 245
Expatriate scientists, 70–71, 282, 283, 299, 303

Filial piety, 151, 153, 163, 165
Fitzgerald, David, 154
Five-Year Plans
  in China, 49
  in Singapore, 66
Fox, Mary Frank, 188
Fraud, research
  in China, 51, 52
  in Japan, 52
  in South Korea, 52
Freedom in research, 107, 124, 232, 233, 235, 296
Fudan University, 44

Gandhi, Rajiv, 59
Gender compromise, 186, 216, 217
  and divorce, 205
  by men, 210
  definition of, 186
  examples of, 200–201, 205, 206, 217
Gender contract, 186
Gender discrimination
  and the career course, 216
  country comparison, 207–208
  in India, 191, 196–197, 212
  in Japan, 208
  in science, 300
  in Singapore, 199
  in the United States, 208
  in the West, 185, 204
  zero-tolerance policy, 300
Gender norms, 204
Gender roles, 213
  in Taiwan, 213
Gender shock, 38, 192
  and Asian male scientists, 185
  and Asian women scientists, 185
  definition of, 184–185, 291
  examples of, 183, 202
  negative, 38, 183, 185, 186, 216, 291
  positive, 185, 202
  upon return, 207
Genome Institute of Singapore, 30
Germany
  scientific culture in, 139, 141, 142
GlaxoSmithKline, 50, 66, 68
Global scientific field, 8
  acquiring status in, 9–10
  actors in, 11
  Asia in, 6, 7, 275, 287
  change in, 8, 287, 288, 302, 303, 305
  core of, 288
  cultural diffusion in, 262
  hierarchy in, 11, 90–92, 181, 248
  networks in, 224
  status in, 168, 180, 227, 246
  status of Asia in, 248
  status of West in, 58
  struggle within, 51
  success in, 216
Google, 296
Grant funding rates
  in China, 51

Index

in the West, 50
Great Leap Forward, 23, 42, 48
Gupta, Namrata, 190, 213

Habitus, scientific
  definition of, 120
Halfway-return migration, 5, 19, 37,
  156, 174, 179, 231, 286, 298
  to Hong Kong, 179
  to Singapore, xvi, 5, 34, 178–180,
    181, 226
Harassment, sexual, 212
Harvard University, 28, 103
  postdoctoral training at, 103
  reputation of, 106
Heroes, national
  scientists as, 180
Hierarchy, importance of, 122, 137,
    277, See also Rank, importance of
  in Asia, 122, 175
  in Japan, 137
  in the United States, 137
  reducing, 278
Higher education sector
  in China, 43–46
  in India, 53–55
  in Singapore, 64
  in Taiwan, 73–74
High-skilled migration, 20
Hiring processes
  in China, 241
  in India, 240
  in Singapore, 240–241
  in Taiwan, 241–242
Hong Kong, 304
  as destination, 304
  as destination choice, 303
  as halfway-return destination, 179
  as origin, 128, 265
  as training hub, 105, 280
  scientific culture in, 140
Hundred Talents Program, xxi, 45, 46,
  157, 170, 175
  Scheme A, 46

Immigration policy
  in Saudi Arabia, xv
  in Singapore, 283
  in the United Kingdom, xiv
  in the United States, xiv, 86

in the West, 93
Imperial College London, 64, 69
India
  and the United Kingdom, 56
  and the United States, 56
  corruption in, 229, 237
  history of, 53
  scientific training in, 61
Indian economy
  liberalization of, 155
Indian Institute of Science, xviii, 28, 29,
    30, 110, 115
  reputation of, 55, 91
Indian Institute of Science Education
    and Research, 4, 55, 270, 298
Indian Institute of Technology, 4, 54,
    55, 94
Information culture, 123, 142
Innovation, scientific
  diffusion from India, 57
  diffusion to China, 48
  diffusion to the West, 47
  in China, 47
  in India, 57
  Western, 48
Institute of Molecular and Cell Biology,
    66
Integration
  definition of, 159
Intellectual property, 294
Interest in the sciences, 265, 269, 284
  by underrepresented groups, 295
  encouraging, 295
  in China, 48
  in high school, 297, 299
  in Singapore, 272
International student migration
  drivers of, 85–88
International student migration system
  historical, 48
  in China, 46, 48
  in India, 56
  in Singapore, 64–65
  in Taiwan, 73–74
International students
  characteristics of, 87
  in Singapore, 64, 265
  in Taiwan, 73, 265
  in Europe, 160–161
  in the United States, 160–161

Internet access
  absence of, 101, 102
  and destination decision-making, 98
  and migration decision-making, 101, 102
  spread of, 98
Intra-Asian mobility, 6, 20
  for postdoctoral training, 251
  for training, 7, 273
  increasing, 287, 301
Ip, Nancy, 189, 190
Ivy League university, 95, 98, 110, 116, 141, 144, 176, 195
  definition of, xxi
  reputation of, 133

Japan, 24, 29, 42, 52, 72, 195
  as destination choice, 43, 48, 105, 137
  as origin country, 26, 30, 100
  as training hub, 105
  gender discrimination in, 183, 185, 189, 197–198
  R&D expenditure in, 75, 169
  return to, 167, 179, 185
  science in, 23, 103, 171, 189
  scientific culture in, 137, 280
  status of, 48
Jia, Hepeng, 50
Johns Hopkins University, 29
Journals, scientific
  cash bonus for publication, 51
  high impact-factor, 51, 52
  Western, 51

Kalam, A. P. J. Abdul, 293
Kuhn, Thomas, 10, 318
Kuomintang, The, 42, 72
Kyoto University, 28, 29

Lancet, The, 10, 51, 83
Lane, David, 68
Latour, Bruno, 120
Leaky pipeline theory, 188, 189, 295, 298–299
Leaning in, 206, 217
Levitt, Peggy, 20, 260, 290
Life course, 16, 185, 207, *See also* Career course
  and career course, 16, 217, 218
  definition of, 16
  impact of, 184
Life sciences, 25, *See also* Biological sciences
Liu, Edison, 68

Mao, Zedong, 42, 43
Masculine compromise, 193, *See also* Gender compromise
Massachusetts Institute of Technology, 28
Maternity leave policies, 300
Matthew Effect, 9
Mazzucato, Mariana, 78
Media images
  and migration decision-making, 97
  of science in Singapore, 68
  of science in the West, 90, 96–98
  of the United States, 97
  of Western universities, 90
Mediascapes, 96, *See also* Media images
Medical migration, xiv, 5
Mentors, xiii, 262
Mentors, scientific, 7, 22, 61, 118, 172, 268, 276, 279, 290
  absence of, 101, 136, 191
  serving as, 279
Merck, 50
Meritocracy, 276
  and diversity, 252, 281, 282, 283
  in China, 50
  in Europe, 140
  in Singapore, 63, 282, 283
  in the United States, 140, 282
Merton, Robert, 10, 21, 22, 58, 119, 139, 144, 276
Migration system, 18
Mode 1 thinking, 70, 231, 296
Mode 2 thinking, 70, 231, 294
Mores of science. *See* Scientific norms

Nanyang Technological University, xviii, 15, 30, 64, 294
  international partnerships, 64
  international ranking of, 64
Nanyang University, 63
National Centre of Biological Sciences, xviii, 30, 59
  reputation of, 122, 169
National Institutes of Health, 76, 231, 247

Index 339

National Science Foundation, 12, 45, 231, 247
  grant application review criteria, 231–232
  grant application review process, 232
National Taiwan University, xviii, 30, 73
National Torch Program, 49
National University of Singapore, xviii, 15, 64, 66, 67, 68, 69, 294
  international partnerships, xii, 64, 68
  international ranking of, 64
Nationalism
  absence of, 174
  and return migration, 171, 172
  and science, 156
  in China, 173
  in India, 156, 172, 173
  in Singapore, 174
  in Taiwan, 173
Nature, 10, 51, 61, 102, 248
Needham, Joseph, 47
Nehru, Jawaharlal, 57, 293
Nepal, 298
Networks, transnational
  student, 90
New England Journal of Medicine, 83
Nobel Prize, 66, 172, 194, 203
Nowotny, Helga, 12, 16

Obama, Barack, 161
OECD, 56
Offshore drug development
  in China, 50
One-China Policy, 78
Opportunity structures
  migration, 90
  training, 90
Optional Practical Training, 86
Organizational culture, 22, 117
Organized skepticism, in science, 21, 119
Oxford University, 29, 116
  reputation of, 108, 274, 276

Pakistan, 194
  as origin country, 30
  gender discrimination in, 204, 208
  gender norms in, 206
  return to, 185, 208

Partnership agreements, interuniversity, 110, 245
  in Singapore, xii, 64, 68, 69
  in Taiwan, 109
  with Canada, 245
Patents Act, India, 60
Patriarchy
  in Asia, 186
  in India, 190
  in Japan, 167, 189
Peking University, 44
Peng, Yinni, 193
Pharmaceutical sector
  in China, 50
  in India, 53, 59–60
  in Singapore, 68, 70
Pioneer Parks, 49
Pipeline, science
  strengthening, 298–299
Policy recommendations, 292–300
Postdoctoral trainees
  availability of, 251
  quality of, 250, 251, 252
  shortage of, 176
Postdoctoral training
  appeal of the US, 88–89
  destination decision, 88–89
  drivers of, 13, 14, 88–89
  in India, 55–56
  increase in, 12
  necessity of, 26
  value of, 15
Problem-solving, approach to, 22, 121, 127–130
  in India, 130
Project 211, xxi, 44
Project 985, xxi, 44

Quality of research
  in Asia, 249
  in China, 44, 51, 79, 172, 275
  in India, 55, 172
R&D expenditure, 4
  of China, 3, 75
  of India, 61
  of Singapore, 296, 297
  of South Korea, 3, 75, 296
  of Taiwan, 75, 296
  of the European Union, 3

R&D expenditure (cont.)
  of the United States, 11
Rank, importance of, 22, 122
  reducing, 277
Research administration, 224, 235–242
  in India, 237, 238, 242
  in Singapore, 239–240
  in Taiwan, 237
  in the United States, 237
  in the West, 115
  inefficient, 235, 242
  professionalization, 295
  purchasing rules, 236–240
Research design, 23–26, See also Research methods
Research funding, 224
  basic vs. applied, 70, 233, 295–296
  importance of, 225
  in China, 48
  in Singapore, 69, 228
  priorities, 233
  white space, 297
Research infrastructure, 225, 253–255
  in Asia, 91
Research methods, 27, See also Research design
Research networks, 96, 224, 249
  absence from, 246–248
  and gender, 247–248
  and location dependence, 96, 167
  building, 243–245
  core of, 243
  in Asia, 96
  losing status in, 95, 113, 167, 246–249
  status in, 224, 243
Research staff, 225, 253
Return migration
  and cultural integration, 154
  and language fluency, 154, 167
  and legal integration, 154
  and nationalism, 156
  and role of familial obligation, 153, 159
  and women, 153
  as a preference, xv
  as selfless, 181
  as senior scientist, 177–178
  axes of influence, 37, 150–152
  decision, 290–291
  incentivizing, 56–57

involuntary, 64, 157, 176–177
location decision, 37, 178
meanings of, 149
negotiation, 164–165
role of familial obligation, 37, 151, 168
role of professional ambition, 37, 151
role of societal integration, 37, 150, 151
timing of, 37
to China, 155, 164–165, 207
to India, xv, xvi, 56–57, 155
to Japan, 207
to Singapore, 176–177, 178
to Taiwan, 77, 132, 155, 173, 208
Robotics, 302
Role model, 243
Role model, female, 202, 203
  absence of, 203
Role of state, 71

Saxenian, AnnaLee, 18
Scandinavia
  scientific culture in, 133, 137
  training in, 132, 137
Schiebinger, Londa, 22, 120
Scholarships, doctoral
  for underrepresented groups, 299
  in Singapore, 65
Science, 10, 102
Science and technology studies, 8, 21, 58, 189, 291
Science education
  in Asia, 33, 84, 125
  in China, 91, 94, 126, 297
  in India, 54, 92, 125, 191
  in Taiwan, 73
  reform, 269–271, 297–298
Science parks, 4, 49, 66, 67, 75, 77
Science policy
  in India, 58
  in Taiwan, 190
Scientific cultures
  as plural, 22, 122
  change in, 7
  defining, 119–121
  definition of, 22
  dimensions of, 121–122
  in Asia, 7, 119–121
  in the West, 36

Index

new, 36
variations across, 7
Scientific knowledge, attitude towards, 22, 121, 123–127
Scientific norms, 21
Scientific remittances, xvi, 7, 19–21, 39, 40, 260
definition of, 260
Scientific research systems
in Asia, 7
variations across, 7
Scientific training
in China, 128
in Japan, 103
Scientific training, in the West
appeal of, 91
Scientist migration system
Asian, 18
Scripps Research Institute, 29
Self-confidence, 130
Serum Institute of India, 53
Shanghai Jiao Tong University, 45
Sharma, Arun K., 190
Shauman, Kimberlee, 188
Singapore
as "Asia-lite," 167, 179
as destination choice, 56
as global city, 181
as halfway-return destination, 179
as Plan B, 251
as stepping stone, 71
as training hub, 105
as welcoming, 265
departure from, 70
foreign scientists in, 71, 174
foreigners in, 180
future challenges, 71
higher education in, 63–64
history of, 62–63
scientific culture in, 280
Social fields
definition of, 9
Social remittances, 20, 290
definition of, 260
of return migrants, 260
Split-household arrangement, 37, 152, 166, 167, 205, 217
Split-household arrangements, 210
Stanford University, 28
reputation of, 106
Star scientists, 96
State science
in India, 57–58
in Singapore, 71
in Taiwan, 75
State, role of, 174, 292–293
in Chinese science, 50
in Singapore, 65
in Taiwan, 76
Stem cell research
in China, 49
in India, 59
Stepwise migration, xi, xii, 87, 101, 105
Stolte-Heiskanen, Veronica, 189
Sunstein, Cass, 106
Superstar scientists, 68, 70, 202, 262, 296
women, 300
Swaminathan, M. S., 59

Taiwan
and the United States, 73–74
as training hub, 280
economy of, 24, 72, 77
geopolitics, 72–73, 79
higher education in, 73–74
history of, 72
out-migration from, 73–74
scientific culture in, 280
Talent war, 17, 23, 46, 234
in Asia, 179, 301, 304
Tata Institute for Fundamental Research, 110
reputation of, 168
Team, research
building, 226
comparing cultures, 128
culture of, 117, 129
diversity of, 274, 280, 281
meetings, 278
productive, 133
recruiting of, 133
women scientists, 191
Technology transfer offices, 294
Tenure policies, 300
Thaler, Richard, 106
Thousand Talents Program, xxii, 46, 157, 262, 282
Tian'anmen Square Massacre, 45

Training hubs. *See also* Singapore, as training hub, Hong Kong, as training hub, Japan, as training hub
in Asia, 105, 109, 298
Translational research
encouraging, 294, 296
in China, 49
in India, 59
in Singapore, 69, 70, 229, 233, 234
in Taiwan, 233
Transnational householding, 155, *See* Split-household arrangement
Transnational migration, 17
from Asia, 18
Tsinghua University, 44, 266, 294
reputation of, 91, 171

UNESCO, 188
United Kingdom, 17
as conservative, 107
as destination choice, 28, 29, 56, 108
centrality of, 105
research funding in, 169
research infrastructure in, 254–255
scientific culture in, 129, 136, 139
training in, 103–104, 107, 124
United States
as a destination choice, 88
as welcoming, 107, 140, 141
return to, 242
scientific superiority, 91
scientific superiority of, 107
Universalism, in science, 21, 22, 58, 71, 119, 139, 140, 277
violation of, 22
Universities
comparing, 122
comparing reputations, 109
in China, 44
in India, 53–55, *See also* Higher education sector, in India
job security in, 15, *See also* Tenure policies
neoliberalization of, xii
public, xii
world-class, xxi, 35, 44, 86, 177
Universities, Asian
ambition of, 110
as public institutions, 236
increasing reputation of, 90–92, 110, 112
reputation of, 93, 99, 102
Universities, private, 28, 116, 170
Universities, Western, 86
corporatization of, 235
scientific superiority of, 114
University of California system, 170
University of California, Berkeley, 28
University of Illinois at Urbana-Champaign, 28
University of Malaya, 63
University of Pennsylvania, 28, 29
University of Science and Technology of China, 45
University of Tokyo, 171
University rankings
increase in, xii
India, 55
Nature Index, 55
Singapore, 64, 110
Taiwan, 73
Times Higher Education Top 100, 29, 55, 64
University restructuring, 68
Upadhya, Carol, 155, 267

Vaccine manufacturing
in China, 52
in India, 53
in Taiwan, 75
Visa, immigrant, xiv, 87
Visa, spousal, 200
Visa, student, 17, 86, 87, 93
Voice of America, 97

Watson, James D., 21
West, The
definition of, 26
technological superiority, 90
Western qualifications
privileging of, 46, 47
Western science
rejecting the concept, 7, 118
Western scientists, 144
in Asia, 303
in Singapore, 283
Women in science
and class privilege, 194

Index

   and parental influence, 194
   awards, 300
   in Asia, 142, 195
   in India, 266
   in Singapore, 191
   in Taiwan, 136
Wong, Joseph, 42
Woolgar, Steve, 120
Work-life balance
   for women scientists, 189
   in Europe, 141
   in Japan, 189

World Health Organization, 73
World Trade Organization, 60, 155

Xiang, Biao, 149, 150
Xie, Yu, 188

Yale University, xxi, 28, 64
   reputation of, 106
Yale-NUS College, xii, 64
Yeo, Philip, 293

Zhejiang University, 29, 45

Lightning Source UK Ltd.
Milton Keynes UK
UKHW021322301121
394829UK00004B/102